Handbook of

CELL AND ORGAN CULTURE

by

DONALD J. MERCHANT
Director
W. Alton Jones Cell Science Center

RAYMOND H. KAHN
Professor of Anatomy
University of Michigan

WILLIAM H. MURPHY, Jr.
Professor of Microbiology
University of Michigan

Burgess Publishing Company
Minneapolis • Minnesota

Library of Congress Catalog Card Number 64 - 22573
Standard Book Number 8087 - 1313 - 2

SECOND EDITION

PREFACE TO THE SECOND EDITION

Development of techniques for the *in vitro* study of animal cells and tissues and the application of these techniques to a wide variety of problems has proceeded at a rapid pace during the past four years. Work has been particularly prolific in virology, radiobiology and cancer chemotherapy, while application has been initiated in animal cell genetics. In revising the *Handbook of Cell and Organ Culture* we have selected important new areas of application for illustrative purposes. The choice of new techniques to be included was made on the basis of their general usefulness to a variety of problems. A number of procedures presented in the first edition have been revised extensively on the basis of these new developments.

To further aid the investigator who wishes to use the book as a laboratory reference manual we have included several standard procedures which often are difficult to locate. The presentation of formulas for media and solutions also has been expanded. Wherever appropriate the formulas presented are those published by the original investigator. The literature citations have been expanded as well as updated and an index has been included to aid the user.

We wish to acknowledge the valuable contributions of many colleagues in the preparation of this volume, particularly: Dr. Richard Eidam, Dr. Raymond Hayes, Dr. Marvis Richardson, Dr. Helene Schneider, Miss Sheila Keefe, Mrs. Fannie Parker, Mr. Thomas Neumeier, Mr. Richard Giles.

Ann Arbor, Michigan
January, 1964

Donald J. Merchant
Raymond H. Kahn
William H. Murphy, Jr.

PREFACE TO THE FIRST EDITION

The application of "tissue" culture techniques to the experimental study of a wide range of biological problems requires an effective means of instruction. Moreover, it is important that students of biology be familiar with the principles of cell, tissue or organ culture in vitro. Since 1953 the Departments of Bacteriology and Anatomy at The University of Michigan have jointly sponsored a course in the "Principles and Techniques of Cell and Organ Culture". This course has been designed to meet the requirements for training at the graduate and post graduate levels. Thus, selection of material for incorporation in this Handbook of Cell and Organ Culture* was made on the basis of considerable experience in teaching students with divergent backgrounds and different interests.

The organization of the exercises is such that they can be used for class instruction or as a guide for the technician or investigator. Introductory statements for each chapter and each exercise have been written with the intent of giving a brief statement of historical development and/or appraisal of the method of procedure. As a further aid, key references have been given to lead the reader into each subject area. In writing the handbook certain opinions had to be expressed and arbitrary decisions made only because of the vastness of the field and the lack of agreement on many matters by respected colleagues.

It is assumed that persons using the handbook have a basic knowledge of microbiological and histological techniques and terminology. To facilitate instruction and to ease the burden of the laboratory assistant, the materials listed for each exercise are those required for one individual to carry out the procedures. Moreover, it is expected that each student is familiar with the technique of sterilizing instruments by immersing them in ethanol followed by ignition. Finally, it is assumed that the laboratory has available the reagents and materials needed for histologic studies of cells.

It is hoped that this handbook will serve as a manual for class instruction and as a guide for the neophyte entering the field of cell and organ culture as well as a useful reference for the more experienced worker.

Ann Arbor, Michigan
January, 1960

Donald J. Merchant
Raymond H. Kahn
William H. Murphy, Jr.

* Supported in part by Grant 2G-195 from the Training Grants Division, USPHS.

TABLE OF CONTENTS

page

Chapter I.
GENERAL LABORATORY INSTRUCTIONS

GLASSWARE

Soft glass (soda-lime glass) is significantly soluble in water and salt solutions. Hard glass, which contains potassium carbonate, is more resistant to solubilization (etching) although this may occur at alkaline pH and at high temperatures. Consequently, exposure of glassware to adverse conditions, such as excessively stringent cleaning procedures or by use of strongly alkaline detergents, will shorten the effective life of laboratory glassware. Soft glass is satisfactory for growth of cells and for short term (1-2 weeks) storage of solutions, although hard glass is superior. For longer storage periods all solutions should be kept in hard glass bottles. Hard glass is marketed under a variety of patented trade names such as Pyrex*, Kimax**, etc. In case of doubt about the qualities of a particular glass product, check with the manufacturer.

Toxicity of glassware is not an uncommon problem which often can be solved effectively by adequate cleaning procedures such as the one described below. In addition, toxicity or its absence is related to the biological properties of the cell strains employed. Unwashed glassware from commercial sources often is not toxic for established cell strains while even clean glassware may be toxic to especially sensitive cells. Detoxification frequently can be accomplished simply by first rinsing glassware with growth medium containing serum or some other biological product rich in protein.

PROCEDURE FOR CLEANING GLASSWARE

Nothing is more basic to cell and organ culture than properly cleaned glassware. Cells often are in direct contact with glass and are maintained with nutrient solutions having low buffering capacity and little or no detoxifying action.

Any of several detergents is satisfactory providing the cleaning and rinsing procedures are adjusted to the properties of the particular compound. Thus cationic, anionic and non-ionic detergents vary in their cleaning effectiveness and particularly in the ease with which they may be removed from

* Corning Glass Company, Corning, N.Y.
** Kimble Glass Company, Vineland, N.J.

glass. In some instances a minimum of 16-18 rinse cycles may be required to reduce the concentration of detergent to a non-inhibitory level though several available detergents require considerably fewer rinse cycles. Glassware washers which embody the principal of sonic agitation may permit use of lower concentrations of detergents and thus minimize these problems. However, the effectiveness of such apparatus is limited by the quantity and type of glassware to be processed.

The following points may be helpful as a general guide:

Immerse all items in a solution of the cleaning agent immediately after use. Large items such as Erlenmeyer flasks, prescription bottles, *etc.* which cannot be handled conveniently in this manner should be kept tightly stoppered until ready for washing. These steps will prevent drying of materials on the glassware and will greatly facilitate subsequent washing. Vessels in which cells have been grown should be brushed before washing. Wax or grease should be removed from slides or other materials with carbon tetrachloride.

Care should be taken to segregate glassware which has been treated with silicone from vessels used for growing cells in monolayers! Minute traces of silicone may inhibit cell attachment and the removal of residual silicone by rinsing procedures is difficult if not impossible. It is good practice to mark each piece of siliconed glassware with an "S" by etching it on the glass with a diamond pencil. Careful segregation of these items in cleaning, sterilization and storage should be adhered to.

It is extremely important to prevent glassware which has been in contact with certain toxic chemicals (formalin, picric acid, colchicine, methylcholanthrene, *etc.*) from being returned to the general glassware stock. In most instances it is not possible or practical to remove such agents from the glass.

Most detergents clean more effectively when used as a warm solution. Boiling generally is not required but may speed the cleaning process and may actually be necessary for some items. To prevent coverglasses from sticking together during washing they should be dropped, one at a time, into a solution of cleaning compound which is boiling vigorously.

Rinsing is the most critical step in the cleaning procedure. All glassware should be rinsed a minimum of 8 times with tap water and twice with distilled water. If an automatic pipette washer is used, 10 cycles with tap water and 2 rinses of distilled water are recommended. Care should be taken to insure that each piece of glassware is filled and emptied completely with each rinse. For all general purposes the distilled water for rinsing may be either singly distilled or deionized water. For critical studies, such as the role of trace metals in nutrition or the action of carcinogenic agents, glass distilled water may be required.

DISPOSABLE PLASTICWARE AND GLASSWARE

Sterile, disposable units including pipettes, tubes, culture dishes and flasks are available currently. Many of these items are fabricated of non-toxic plastics and the plastic surfaces permit the growth of most tissue cells in a highly satisfactory manner. Optical quality of plastic flasks often exceeds that of glass vessels of comparable price. A number of inexpensive, disposable glass products are also marketed including tubes and pipettes.

The variety of plastic or glass disposable items is being expanded rapidly. Considerations other than availability in selecting plastic versus glassware are: a) comparative cost (including cleaning of ordinary glassware and replacement), b) avoidance of toxicity due to chemicals and detergents when using disposable items, c) limitations of plastics relative to their intrinsic physical and chemical properties. (In the latter category is a rather severe limitation on use of common fixatives and staining reagents).

ASEPTIC TECHNIQUE

The design of an aseptic technique depends on the needs of the individual and the demands of his problem. Although it is advantageous to have sterile transfer rooms or hoods, they are not required for many cell and organ culture procedures. In that microorganisms are most commonly airborne on dust particles, it is essential that work areas be kept as free of dust as possible. Air turbulence should be avoided and for this reason the transfer area should be situated away from major traffic patterns.

Whenever it is feasible the use of screw-cap vessels or of caps which fit well over the mouth of the vessel (*e.g.* , Morton Closures^R*) will aid in maintaining sterility. This is particularly important when it is necessary to decant from one container to another.

USE OF ANTIBIOTICS

The availability of antibiotics selectively toxic for many microbial agents has aided materially in the development and application of cell and organ culture techniques. The use of these materials in large scale tissue culture work or in the presence of known contaminated materials is certainly justified although the routine use of antibiotics in culture work is not without limitations and hazards. The decision to use antibiotics in tissue culture studies should be made only after a careful evaluation of the necessity and of the consequences.

* Bellco Glass Co., Vineland, N.J.

The criteria most commonly used to determine toxicity of antibiotics are gross changes in morphology or growth patterns. It is quite clear that more subtle but significant effects may be missed. Thus, the current feeling that the commonly used antibiotics have little or no effect on tissue cells may not be justified. Additionally, the apparent lack of toxicity for one or more cell strains in a test series may have little meaning when extrapolated to other cell strains and/or other experimental conditions.

A serious problem relates to the suggestion that routine use of antibiotics is responsible for the widespread occurrence of PPLO or L-forms of bacteria in tissue culture cell lines (1). The knowledge that penicillin may be used to establish L-forms from bacteria (2) is reason enough to suggest caution. Continued use of antibiotics also may mask bacterial or fungal infections and lead to a chronic or latent type of infection.

When the use of antibiotics is deemed necessary the probable type of contamination which may be encountered will determine which antibiotics to use. To control the widest range of bacterial contaminants it is common to employ either a broad spectrum antibiotic such as neomycin or a tetracyline or to use a combination of penicillin and streptomycin (p. 242). The latter are far less toxic but penicillin, as mentioned above, may induce L-forms. Though the common antibiotics effective against bacteria have little or no activity against the common fungi several antifungal agents are available. Among these nystatin and amphotericin B have demonstrated value.

Materials known to be contaminated and which must be added to cell culture systems, such as clinical specimens for virus isolation, are often treated with high concentrations of antibiotics for limited periods prior to use.

Once cultures become contaminated it is almost useless to attempt a "cure" with antibiotics. Time and effort are generally wasted. The infection may either be partially suppressed and break through as soon as antibiotics are removed or stable L-forms may be induced. An exception is the use of tetracycline or kanamycin to free cell lines of PPLO or L-forms (3, 4). This appears to be effective, at least in certain cases.

The most commonly employed antibiotic combination for control of bacterial contamination is penicillin (50 units/ml) and streptomycin (50 μg/ml) with or without the addition of nystatin (30 μg/ml). Other antibiotics frequently used are tetracycline (5 μg/ml), neomycin (25 μg/ml), kanamycin (50 μg/ml) and amphotericin B (2.5 μg/ml). It is to be emphasized that these antibiotics and particularly the dosages listed are only examples and are not recommended necessarily for any particular situation. For serum containing media it will often be necessary to approximately double these amounts. To rid cultures of PPLO or L-forms, much higher concentrations of tetracyline and kanamycin are required.

STERILIZATION PROCEDURES

Autoclaving, dry heat sterilization and filtration as they are commonly used in microbiology are adequate for most cell and organ culture materials. A few suggestions with regard to choice of technique and points of procedure are given below.

Autoclaving is the method of choice for most solutions and is often used for sterilization of glassware, particularly when heat penetration may be a problem. For some types of critical work it may be necessary to know the quality of steam supplied to the autoclave. Depending on water conditions, various compounds are added to boilers to minimize rusting. Certain of these materials volatilize and become a source of toxic substances deposited within the autoclave. A filter unit is available* which can be placed in the steam line to avoid this difficulty. A more common and serious hazard is failure to sterilize due to improper loading of the autoclave. Care should be taken to avoid overcrowding of materials.

A temperature of 121°C applied for 20 minutes is recommended for most materials. If all air in the autoclave has been replaced by steam this temperature will be achieved at a pressure of 15 lbs/sq. in. It is imperative that timing be commenced _only when the temperature has reached 121°C_! For materials such as glucose, which are unstable at high temperatures, the temperature may be lowered providing the time interval is increased accordingly. In the case of steam sterilization of _screw cap containers_ it is imperative that the _caps be left quite loose until after the vessels have been removed from the autoclave and allowed to cool_! This will prevent production of a vacuum which can result in pulling contaminated air into the vessels as they are opened for use.

Dry heat sterilization is effective for glassware unless rubber or teflon-lined caps or other rubber or plastic parts are involved. Heat transfer is slow and the timing of sterilization must be made from the time the material in the oven reaches the required temperature. Again, proper loading of the oven is essential. A minimum of 170°C for 2 hours is necessary. If a manually operated gas oven is used, 190°C for 30 minutes is adequate.

Filters may be used to remove bacteria, yeasts and molds from solutions but do not remove most viruses or PPLO and L-forms of bacteria. Four types of filters are commonly employed for this purpose, namely, asbestos pads, membrane filters, sintered glass and unglazed porcelain. _Mean effective pore diameter_ of the filter should be 0.5 μ or less. Effective filtration cannot be attained if the filter is allowed to become saturated with organisms.

* Selas Corporation, Dresher, Pa.

Therefore, the amount of a material which can be filtered safely will depend upon the numbers of organisms it contains and the volume of material to be filtered.

Viscous or protein containing solutions such as serum generally require pressure filtration. This speeds up the filtration process and also prevents foaming. In using pressure, however, it is necessary to determine the maximum safe pressure for the filter. If large amounts of solids must be removed from solution, sedimentation by high speed centrifugation or a prior filtration through a clarifying type unit may be required. Rate of filtration generally is not an important factor as long as the integrity of the filter is maintained. With membrane or asbestos filters, clarifying pads may be inserted in series before the sterilizing pad to speed the filtration procedure. For small amounts of material a Swinney adapter* which fits any standard hypodermic syringe is quite advantageous. Both asbestos and membrane filter pads are available to fit these adapters.

Asbestos and membrane filters are expendable, thus eliminating the problem of cleaning. Asbestos pads or membrane filters may develop leaks if not used properly and asbestos pads also may contribute inorganic ions, particularly Mg^{++}, to the solution being filtered if not previously washed. This is generally an important consideration only when small volumes are being filtered. With small volumes an appreciable amount of material may be lost due to absorption to the pad. Membrane filters generally have a more rapid flow rate than asbestos pads unless very viscous or turbid suspensions are being filtered. In the latter cases clarifying prefilters are useful. In contrast to asbestos pads, absorption on the filter and contribution of inorganic ions are minimized with membrane filters.

Unglazed porcelain filters should be checked for cracks or imperfections by measuring bubbling pressure. This is done by placing the filter element in a cylinder containing distilled water and connecting the filter to a sensitive compressed air source. Selas 02 filters** have a bubbling pressure of 25 lbs/ in^2 while the 03 filter** has a bubbling pressure of 35 lbs/in^2. If air bubbles appear before these pressures are attained a crack or flaw in the filter is indicated. Cleaning of a porcelain filter is done by rinsing first with 2% $NaHCO_3$ or 2% NaCL to remove residual proteins. The filter candle is then soaked overnight in concentrated HNO_3 or dichromate solution, rinsed thoroughly, dried at $110^{o}C$ and then heated in a muffle furnace with a rise of $160^{o}C/hr.$ until the temperature reaches $675^{o}C$. Hold the filter for one hour at this temperature, and then allow it to cool slowly to room temperature inside the furnace. Reverse flush the filter with distilled water to remove ash, soak it in distilled water several hours and check bubbling pressure as indicated above. Sintered glass filters are best cleaned with hot sulfuric or nitric acid to remove organic materials. Reverse flush the filter but avoid using pressure in excess of 15 lbs/in^2. Wash them thoroughly in distilled

* Becton, Dickinson, Co., Rutherford, N.J.
** Selas Corporation, Dresher, Pa.

water and rinse with ethanol and acetone to remove water. With both glass and porcelain filters care should be taken to avoid sudden temperature rise if the filter is damp as production of steam may crack the filter.

TESTING FOR STERILITY

All media and solutions to be used in cell or organ culture work must be carefully pretested for sterility utilizing a representative sample. On the other hand it is equally important to recognize the limitations of sterility tests. Few, if any, test media are as sensitive growth indicators as the tissue cultures themselves. Moreover, due to statistically predictable difficulties involved in obtaining an adequate sample, one cannot be certain that a particular batch of medium is sterile unless the entire batch is consumed in testing. A practical rule of thumb is to take 1% of the specimen for a sterility test.

Routine testing procedures should be designed to include the normal range of aerobic and facultatively anaerobic bacteria as well as yeasts and molds. In view of the evidence of PPLO and/or bacterial L-forms in cell strains, tests for their presence also should be made at frequent intervals. If serum or other natural products are used they should be checked for the presence of viruses.

For routine detection of bacteria both brain heart infusion broth*, ** (or trypticase soy broth*) and fluid thioglycollate*, **, *** should be inoculated in duplicate. One tube of each medium should be incubated at room temperature and the other set at 35°C for a minimum of one week. Where yeasts and molds are particularly troublesome it will be advisable to repeat the above procedure with an acid glucose broth and plates of Sabouraud's agar*, **. In cases where contamination cannot be recognized readily in the sterility tests, serial passage may be helpful. To increase the chances of detecting contaminants in tissue culture medium the entire batch may be incubated overnight and then tested for sterility. It is wise to make a Gram stain of any contaminants which are isolated and to keep a record of the types of organisms found. This may be helpful later in tracing the source of contamination.

To test for presence of L-forms of PPLO, plate 0.1 ml of material on the surface of each of several plates of PPLO agar (5) in 4 cm petri dishes. If cell cultures are being tested, it is advisable to use the medium from a heavy monolayer or from the plateau phase of growth of suspension cultures. If L-forms or PPLO are present in such cell cultures they will be in the

* Baltimore Biological Laboratories, Baltimore, Md.
** Difco Laboratories, Detroit, Michigan
*** Should not contain methylene blue which is toxic to some organisms.

supernatant medium in numbers which can be detected. Cultures must be incubated in 5% CO_2 and 95% N_2 at 35°C (or 37°C). After 5-7 days examine the surface of the plates with a hand lens or a dissecting microscope for the presence of colonies, which range in size from 50-300 μ, and which characteristically have a "fried-egg" appearance. Staining of the colonies with methylene blue-azure is an aid in identification (6). For confirmation, a colony may be dug from the agar, crushed on a slide and observed with the phase contrast microscope.

Presence of viruses in serum, ascites fluid, *etc.* may be checked by testing the materials on one or two cell strains of known broad sensitivity, *e.g.* , rhesus monkey kidney cells. Evidence of cytopathology, as compared to control cells, is taken as presumptive evidence of a viral agent (or primary toxicity). Specimens showing such activity should be discarded. For routine purposes it is not worthwhile to attempt the identification of an agent or to differentiate between toxicity and a viral agent.

PREPARATION OF MEDIA AND SOLUTIONS

Chemicals of at least reagent grade should be used for preparation of media and solutions. For careful nutritional studies and when used to prepare defined media it will be necessary to rely on suppliers who can verify composition of their products. For many purposes deionized water will serve quite adequately. However, it should be recognized: (a) that many organic compounds are not removed by ion exchange resins, (b) that the quality of the water used to charge the resin will affect both the quality of water produced and the effective "life" of the resin and (c) certain bacteria may grow in the resin and contribute toxic materials to the water. Glass distilled water is safer to use than deionized water and several stills are available commercially which provide a high quality product at a rate which will meet the needs of most laboratories.

A wide range of media and medium components are now available from the sources listed in the appendix. The choice as whether to purchase such materials or to prepare them from natural sources or from stock components must be decided by each individual worker. Factors which will influence this choice in any particular instance will include relative cost (including preparation), availability, facilities of the laboratory and the degree to which the individual must control his particular system.

pH CONTROL

In most cell or organ culture systems pH control is provided, in part, by the carbonate-bicarbonate system. To maintain the proper pH level with such a system it is necessary to use tightly stoppered vessels or to gas with 5% CO_2. In most cases the closed system is adequate. pH control is particularly difficult when chemically defined media are used since the added buffering capacity of serum or other proteins is not available.

Adjustment of pH of solutions and media may be accomplished, with maintenance of isotonicity, by use of 0.3 N NaOH and 0.3 N HCl. Excessive use of HCl, however, will drive off CO_2. pH may also be raised by the addition of isotonic $NaHCO_3$ (1.4%) or lowered by equilibration with gaseous CO_2. It is advisable to add phenol red to all solutions in a concentration of 0.01-0.02 grams/liter to facilitate approximation of pH.

MAINTENANCE OF STOCK CULTURES

Cell lines which are routinely used are maintained most conveniently by serial subculture. The frequency of transfer will be determined largely by the growth rate of the particular culture, the medium used and the temperature of incubation. To obviate the necessity of frequent handling the cells can be carried on a medium which is suboptimal for rapid proliferation. Another strategem consists of incubating cells at lower temperature (usually 28°-30°C) to prolong the generation time. In either case care must be taken not to select the population for ability to grow under the suboptimal conditions.

If cultures are carried serially in suspension culture, it is advisable also to maintain stocks of the cells in monolayer to permit microscopic examination. Stock lines should always be carried without the routine addition of antibiotics. At monthly intervals each stock line should be checked to make certain that it is free of bacterial, fungal, PPLO and viral contamination. It is suggested also that the chromosomal complement (idiogram) (p. 198) and species antigen (p. 114) be checked at least twice a year.

For a discussion of long term storage of cell lines, see page 207.

REFERENCES

1. ROTHBLAT, G. H., and MORTON, H. E., Detection and possible source of contaminating pleuropneumonialike organisms (PPLO) in cultures of tissue cells, Proc. Soc. Exper. Biol. & Med., 100: 87-90 (1959).
2. KLIENEBERGER-NOBEL, E., Origin, development and significance of L-forms in bacterial cultures, J. Gen. Microbiol., 3: 434-443 (1949).
3. HEARN, H. J., JR., OFFICER, J. E., et al., Detection, elimination, and prevention of contamination of cell cultures with pleuropneumonialike organisms, J. Bact. 78: 575-582 (1958).
4. FOGH, J., and HACKER, C., Elimination of pleuropneumonialike organisms from cell cultures, Exper. Cell Res., 21: 242-244 (1960).
5. BARILE, M. F., YAGUCHI, R., et al., A simplified medium for the cultivation of pleuropneumonialike organisms and the L-forms of bacteria, Am. J. Clin. Path., 30: 171-176 (1958).
6. DIENES, L., Morphology and nature of the pleuropneumonia group of organisms, J. Bact., 50: 441-458 (1945).

Chapter II.
CELL CULTURE TECHNIQUES

GROWTH OF ESTABLISHED CELL LINES

EXERCISE 1. MONOLAYER CULTURES

The technique of cultivating animal cells on a solid surface using a fluid overlay was originated and developed by Earle and co-workers (1-5). The method is adaptable to the growth of a wide spectrum of established cell lines and of primary dispersed tissue cells. It has the advantage of being simple and reproducible and at the same time permits detailed microscopic study of cells when they are grown on coverglasses.

In the monolayer technique cells attach to the glass, or other surfaces, flatten and grow to yield a continuous sheet, or network, usually one cell thick. Cells in such a monolayer may be spindle shaped, polygonal or amoeboid (p. 175). When crowding occurs the cell layer may loosen from the surface and slough off.

Cells are harvested from monolayer culture by scraping, by treating with a chelating agent such as ethylenediaminetetraacetic acid, or by treatment with an enzyme solution such as trypsin, pancreatin, or collagenase. The choice of harvesting procedure will depend, in some measure, on the cell strain being used and the composition of the medium.

MATERIALS

1. Monolayer cultures of L-M strain of mouse cells (6) and HeLa strain of human cervical carcinoma cells (7).
2. Thirty ml sterile 199 peptone (199P) (p. 220).
3. Thirty ml sterile $Eagle_{80}$ human $serum_{20}$ (p. 237).
4. Ten ml sterile 0.05% trypsin or 0.02% EDTA (p. 240-241).
5. Ten ml sterile, CMF-PBS solution (p. 240).
6. Twelve sterile, cotton plugged 10 ml serological pipettes.
7. Four sterile, cotton plugged 1 ml serological pipettes.
8. Ten sterile 2 oz French square bottles with rubber lined screw caps.
9. One sterile 15 ml screw-cap centrifuge tube.

10. Twelve sterile screw-cap 16x125 mm tubes.
11. Twelve sterile screw-cap 16x125 mm tubes (or Leighton tubes*) each containing one 9x22 mm #1 coverglass.
12. Sterile rubber policeman.
13. Haemocytometer.
14. Ten ml citric acid-crystal violet diluting fluid (p. 242).

PROCEDURE

1. Examine stock cultures microscopically and become familiar with the patterns of growth.

2. Harvest the cells by the following procedures:

   a. Remove the medium from the culture of L-M cells and replace with 10 ml of fresh 199P. Scrape the cells from the glass surface with a rubber policeman. Be certain to remove all cells. Triturate the suspension by pipetting 3-4 times with a 10 ml serological pipette by drawing up the cells and forcibly expelling them.

   b. Remove the medium from the culture of HeLa cells and wash once with an equal quantity of CMF-PBS solution. Replace the CMF-PBS with 10 ml of 0.05% trypsin or 0.02% EDTA. Incubate at room temperature until the cells detach from the glass. Triturate and transfer to a 15 ml centrifuge tube. Centrifuge at 500 RPM for 5 min. Decant the trypsin or EDTA, resuspend the cells in 10 ml of $Eagle_{80}$ human $serum_{20}$ and triturate.

3. Count the cells (from 2a or 2b above) in a haemocytometer using citric acid-crystal violet solution (p. 155) Dilute the suspension to 2×10^5/ml for the L-M strain and 1×10^5/ml for HeLa strain using the respective media.

4. Each student will prepare 5 bottle cultures, 6 tube cultures and 6 coverglass cultures of <u>each</u> of the cell strains as follows:

   a. Dispense 3 ml aliquots into each 2 oz French square bottle and 1 ml into each culture tube (with and without coverglasses).

   b. To prevent loss of CO_2 be certain that all vessels are tightly stoppered. (Be sure screw-caps have rubber liners.)

   c. The final pH should be 7.4-7.6 (p. 8).

* Bellco Glass Co., Vineland, N.J.

NOTES

5. Incubate at 35°C. Tubes should be placed in racks which permit them to be tilted at a suitable angle. Tubes containing coverslips preferably should be incubated in a horizontal position.

6. Examine the cultures daily and adjust the pH or change medium as indicated:

   a. In the case of L-M strain the pH will not need to be adjusted. A medium change at 5 days is adequate. The cells should form a dense monolayer in 7-9 days.

   b. HeLa cells will probably require a medium change each 2-3 days. Adjustment of pH may or may not be necessary between medium changes (p. 8) but pH should not be allowed to drop below 7.0. A dense monolayer will be formed within 7-9 days.

7. Harvest one bottle of each cell strain on days 3, 4, 5, 6, 7. Suspend and disperse, make a total cell count and plot log to base 2 (p. 253) of cell number against time in days (8). Cytologic examination may be accomplished by the use of the collodion stripping technique (p. 179).

8. Harvest one coverglass culture of each strain on days 2, 3, 4, 5, 6, 7. Remove the coverglasses from the tubes, fix in FAA (p. 177) and stain with haemotoxylin and eosin (p. 185).

NOTES

EXERCISE 2. PROPAGATION OF CELLS IN SUSPENSION

Growth of mammalian cells in suspension was demonstrated in 1953 by Owens, Gey and Gey (9) using a strain of mouse lymphoblastoma. Earle and co-workers (10) showed the method to be applicable for growth of L strain fibroblasts and later for other established cell lines. Large scale growth of cells in suspension has been achieved by McLimans and others (11-13).

Because of the size of animal tissue cells (approximately 16 μ in diameter) some method of agitation is required to keep them suspended in a liquid medium. Owens, *et al.* (9) accomplished this with a "tumble tube" which is suitable for small scale studies. Growth of cells in larger volumes to permit repeated sampling and the use of quantitative methods is readily accomplished by use of a rotary shaker or by stirring. The shaker technique was introduced by Earle, *et al.* (10) and is particularly suited to studies requiring replicate cultures. The spinner culture has the advantage of being more versatile by virtue of permitting variations in speed of rotation, volume, *etc.* This becomes particularly important if more than one or two cell lines are cultivated by this method since the optimum conditions for proliferation in suspension may vary appreciably from one cell line to another. Multiple spinner assemblies are now available which make possible replicate studies by the spinner method.*

The suspension culture method is particularly well suited to quantitative studies of nutrition, metabolism, population analysis, *etc.* By increasing the scale of operation it is possible to produce large amounts of cells for analysis. Some strains grow in suspension to give monodisperse cultures (14). Others grow only as small clumps or clusters of cells.

SHAKER CULTURE METHOD

MATERIALS

1. Monolayer cultures of L-M strain mouse cells.
2. One hundred ml sterile 199 peptone (199P) (p. 220).
3. Sterile 4% solution of 15 cps. methylcellulose (p. 241).
4. Sterile 250 ml Erlenmeyer flask with rubber lined screw cap.
5. Two sterile cotton plugged 1 ml serological pipettes.
6. Four sterile cotton plugged 10 ml serological pipettes.
7. Sterile 4 oz graduated prescription bottle.
8. Haemocytometer or electronic cell counter (p. 155-156).
9. Citric acid-crystal violet diluting fluid (p. 242).
10. Four tenths percent solution of erythrosin B (p. 157).

* Eberbach Corporation, Ann Arbor, Michigan.

NOTES

PROCEDURE

1. Replace the medium on a monolayer culture of L-M strain fibroblasts with 10 ml of fresh 199P.

2. Scrape the cells into the medium with a rubber policeman and triturate with a 10 ml pipette. Determine the number of cells per ml (p. 155-156).

3. Pipette 2 x 10^7 cells into a graduated 4 oz prescription bottle and dilute to 100 ml with 199 P containing 0.12% methylcellulose (3 ml of 4% solution per 100 ml medium). Transfer the cell suspension to a 250 ml Erlenmeyer flask.

4. Close the flask tightly and place on a rotary shaker at 110 RPM and 35°C.

5. Each day withdraw a 1.0 ml sample and make a total count (p. 155-156) and a viable count (p. 157). Plot the log to base 2 of cell number against time in days (p. 253).

SPINNER CULTURE METHOD

MATERIALS

1. Suspension culture of L-M strain mouse cells in late logarithmic growth phase (from shaker culture above).
2. One sterile 250 ml spinner flask.
3. One hundred ml of 199 Peptone (199P) (p. 220) in 4 oz prescription bottle.
4. One sterile cotton plugged 1 ml serological pipette.
5. One sterile cotton plugged 100 ml volumetric pipette.
6. Haemocytometer or electronic cell counter (p. 155-156).
7. Sterile 4% solution of 15 CPS methylcellulose (p. 241).
8. Citric acid-crystal violet diluting fluid (p. 242).
9. Four tenths percent solution of erythrosin B (p. 157).
10. Three sterile cotton plugged 10 ml serological pipettes.
11. Two sterile 15 ml conical screw-cap centrifuge tubes.

PROCEDURE

1. Determine the number of cells per ml (p. 155-156) in the suspension culture (from shaker culture above).

NOTES

2. With the sterile 10 ml pipette transfer an aliquot of the cell suspension containing $2x10^7$ cells to the sterile centrifuge tubes. Centrifuge at 500 RPM for 10 minutes.

3. Decant the supernatant medium from the sedimented samples and resuspend the cells from both tubes in the 100 ml of 199P. Add 3 ml of 4% methylcellulose and mix well. Determine the number of cells per ml (p. 155-156). It should approximate $2x10^5$/ml.

4. With the sterile 100 ml volumetric pipette transfer 100 ml of cell suspension into the 250 ml spinner flask. Tighten the closures.

5. Place the spinner flask in the water bath of the multiple spinner assembly at 300 RPM and 35°C.

6. Each day withdraw 1 ml of sample and make a total cell count as well as a viable count (p. 155-157). Plot the log to the base 2 of cell number against time in days (p. 253).

NOTES

EXERCISE 3. CULTIVATION OF CELLS ON SOLID MEDIA

Very early studies by the Lewises (15), Weil (16) and others suggested that growth of cells on agar solidified media might be a useful procedure. Later Zinsser, *et al.* (17), Cheever (18) and others applied the method to the cultivation of viruses and rickettsia. Wallace and Hanks (19) have demonstrated growth of several cell lines and primary tissues on media solidified with agar.

Maintenance of viability for prolonged periods (6 weeks or longer), simplicity of technique, and ease of handling and storage, as well as other factors suggest that this may prove to be a very useful method for carrying stock cultures as well as for experimental studies.

MATERIALS

1. Monolayer culture of L-M strain of mouse cells.
2. Monolayer culture of HeLa cells.
3. Twenty-five ml sterile 2X 199 peptone (p. 238).
4. Five ml sterile 1X 199 peptone (p. 238).
5. Twenty-five ml sterile 3% agar in distilled water.
6. Ten sterile 16 x 125 mm tubes with rubber lined screw caps.
7. Bacteriologic inoculating loop.
8. Sterile 15 ml centrifuge tube with rubber lined screw cap.
9. Dessicator with candle or gassing chamber with tank of 5% CO_2 in air.

PROCEDURE

1. Melt the 3% agar and cool to 50° C. Warm the 2X 199 peptone to the same temperature, mix the two in equal proportions and distribute 5 ml per screw-cap culture tube. Allow to solidify to form a long slant. Add approximately 0. 3 ml of 1X 199 peptone (fluid) to the base of each slant.

2. Harvest cells as directed in Exercise 1, steps 2a, b (p. 12).

3. Centrifuge the cell suspension at 165 G for 5 min. Decant the supernatant fluid as completely as possible and resuspend the cells in the small volume of fluid remaining.

4. With an inoculating loop transfer a loopful of the dense suspension of cells to the surface of each of 10 slants. Touch the loop to several spots on the surface of the slant.

NOTES

5. Incubate the tubes at 35°C in an upright position in a dessicator or other tightly closed container. Light a candle in the container just before it is closed. This will give an atmosphere containing approximately 5% CO_2 but with a reduced O_2 tension. An alternative method is to use a chamber in which the atmosphere can be replaced with 5% CO_2 in air. The caps on the tubes must be loose.

6. Examine at 2-3 day intervals for the presence of colonies and note their characteristics: size, consistency, color, *etc.*

7. At weekly intervals (for 6 weeks) transplant to a fresh agar medium using an inoculating loop. The inoculum should be taken from the edge of the colony. At the same time test for viability (p. 157).

8. After 10-12 days remove a colony from the agar and fix *in toto* in formalin-acetic acid-alcohol (p. 177). In removing the colony it is desirable to cut out a block of agar containing the colony to prevent possible distortion. Section and stain with haemotoxylin and eosin (p. 185). Examine for characteristics of colonial growth.

9. After 1-2 weeks harvest a slant, wash off the cells with 1X 199 peptone and replant as a monolayer on coverslips as described in Exercise 1. Stain and compare with the cells which have been grown as a monolayer on glass.

NOTES

PRIMARY CULTURES FROM DISPERSED TISSUES

Primary cell cultures differ from "established cell lines" because they are not propagated serially *in vitro* and are comprised of mixed cell populations which may be relatively heterogeneous in morphology and in other biological properties. Primary cultures of trypsin or EDTA dispersed cells can be used advantageously because: (a) the expense and inconvenience of maintaining established cell stocks is obviated; (b) they are particularly suitable for vaccine production since the probability of *in vitro* transformation of cells to malignancy is minimized; (c) massive quantities of tissue can be obtained conveniently for short term studies; (d) they are hardy and can be sustained well in media of relatively simple composition; and (e) their degree and range of sensitivity to viral infection may exceed that of the common established cell lines. The use of primary cell cultures also has inherent disadvantages: (a) they may be contaminated with latent viruses (*e.g.* , foamy virus in monkey kidney tissues); (b) long term experiments concerning the biological properties of cells cannot be carried out; and (c) mixed populations of cells may confuse interpretation of experimental findings.

Primary cell cultures can be prepared from tissues that are relatively inexpensive and convenient to obtain. For example, human amnion cells (20-24) and primary cell cultures of renal tissues from animals are satisfactory for detection and assay of a variety of human and animal viruses. The following exercises review the procedures for the preparation and cultivation of primary cell cultures.

NOTES

EXERCISE 1. PREPARATION OF MONOLAYER CULTURES FROM RABBIT KIDNEY

This method is essentially the same as that used for cultivation of monkey renal cells (25).

MATERIALS

1. Sterile surgical instruments: scalpel, mouse tooth forceps, iris scissors, Bard-Parker blades and handles.
2. Five 1 ml and fifteen 10 ml sterile, cotton plugged, serological pipettes.
3. Two sterile Petri dishes.
4. Sterile Seitz filter apparatus equipped with a double thickness of cheese cloth as a substitute for the Seitz filter pad.
5. Sterile screw-cap "trypsinization-flask"* containing sterile glass beads and a teflon-covered magnetic stirring bar.
6. Magnetic stirring apparatus.
7. Twenty sterile 2 oz French square bottles with rubber lined screw caps.
8. Haemocytometer.
9. Sterile CMF-PBS (p. 240) containing penicillin and streptomycin in final concentrations of 100 units and 100 μg per ml respectively.
10. Sterile trypsin solution, 0.25% in CMF-PBS (p. 240).
11. Ten sterile screw-cap, 15 ml conical centrifuge tubes.
12. Growth medium of the following composition:

Lactalbumin hydrolysate**	0.5 gm
Calf serum	2.0 ml
Sodium bicarbonate solution (1.4%)	2.5 ml
Hanks BSS (balanced salt solution)	95.0 ml

(Prepare the BSS so that it will contain penicillin and streptomycin in the concentrations used in the CMF-PBS described in 9 above. NOTE: DO NOT add the calf serum to the medium until the cells have been enumerated as described below. Calf serum causes rabbit renal cells to agglutinate.)

13. One rabbit, 6 to 8 weeks of age. (Older rabbits are not satisfactory).

PROCEDURE

1. Exsanguinate a rabbit. Remove and transfer the kidneys aseptically to a sterile Petri dish. Wash the kidneys with CMF-PBS (p. 240) to remove extraneous materials.

* Bellco Glass Co., Vineland, N.J.
** Nutritional Biochemicals, Cleveland, Ohio.

NOTES

2. Use a sterile fine-tipped forceps to strip the capsules from the kidneys; discard the capsules. Separate by means of iris scissors the cortical (outer) region from the medullary (inner) region; discard the medullary (light) tissue.

3. Mince the cortical tissue with scalpel blades until 1 mm^3 pieces are obtained. Wash the minced tissue three times with cold CMF-PBS; discard the washings.

4. Transfer the minced tissue to the trypsinization flask.

5. Add sterile stock trypsin solution (0.25%) chilled to 4° C to the trypsinization flask and incubate the tissue at 4° C with constant stirring for 6 to 7 hours. After incubation centrifuge the cells, discard the supernatant fluid which contains toxic materials, and add fresh cold trypsin. Continue trypsinization for an additional 18 hours.

6. Separate fibrous tissues from the trypsinized cells by filtration through the cheese cloth contained in the Seitz filter apparatus.

7. Sediment the cell suspension and discard the supernatant fluid. Wash cells three times with cold (4°C) CMF-PBS.

8. Centrifuge the washed cells at about 600 RPM for 2 minutes to separate epithelial cells (sediment) from the erythrocytes contained in the supernatant fluid. The critical factor in this step of the procedure is to separate erythrocytes from epithelial cells; the velocity and time of centrifugation can be modified to provide the desired results.

9. Suspend the cells in growth medium free of calf serum (to prevent agglutination). Take an aliquot of cells (aseptically) and dilute in CMF-PBS to give a suspension countable in a haemocytometer (p. 155).

10. Add calf serum to the growth medium and dilute the cell suspension with this complete growth medium so that 10^6 cells will be delivered in a volume of 3 ml into each of twenty 2 oz French square bottles.

11. Incubate cultures at 35°C; feed with growth medium as needed. Cells that appear degenerated will frequently grow and therefore should not be discarded prematurely.

NOTE: Primary cultures can be trypsinized to give secondary cultures but further subcultures usually fail to grow. Cell multiplication occurs in secondary cultures; moreover, the cells have excellent morphologic characteristics.

NOTES

EXERCISE 2. PREPARATION OF CULTURES FROM HUMAN AMNION

Either a rapid digestion procedure at 35°C or a slow procedure at 4°C may be employed depending upon the needs and the routine of the laboratory.

MATERIALS

1. One full-term human placenta in CMF-PBS (p. 240) containing 200 units of penicillin and 200 μg of streptomycin per ml.
2. Four sterile Petri dishes.
3. Two sterile 250 ml screw-cap Erlenmeyer flasks each containing a teflon covered magnetic stirring bar or a sterile trypsinization flask (see exercise 1, p. 28).
4. One magnetic stirrer.
5. One sterile cotton plugged large orifice 50 ml volumetric transfer pipette.
6. Sterile surgical instruments: iris scissors, two #11 Bard-Parker blades with handles.
7. Twelve sterile 16x150 mm culture tubes with rubber lined screw caps.
8. Six sterile screw-cap Leighton tubes containing 9x22 mm cover-glasses.
9. Four sterile 2 oz. French square bottles with rubber lined screw caps.
10. Sterile 10 cm funnel containing a gauze pad of 4 thicknesses. Top and stem covered with aluminum foil.
11. One liter CMF-PBS (p. 240) containing 200 units penicillin and 200 μg streptomycin/ml.
12. Five hundred ml of 0.25% trypsin* (or pancreatin) (26) in CMF-PBS containing 200 units penicillin and 200 μg streptomycin per/ml.
13. One hundred ml of Eagle medium plus 20% human serum and containing 100 units of penicillin and 100 μg of streptomycin/ml.
14. Four sterile screw-cap 150 ml conical centrifuge bottles.

PROCEDURE

1. Carefully separate the amniotic membrane of the placenta from the chorion (villous membrane) and place the amnion in a sterile Petri dish.

2. Add 10-15 ml of trypsin (or pancreatin) solution to the membrane and incubate for 15-20 minutes at 35°C.

* Difco Laboratories, Detroit, Michigan (Difco 1:250)

NOTES

3. Aspirate and expel the trypsin solution several times with a ten ml pipette and then remove. This preliminary treatment should remove most of the blood clots. Add 10 ml of fresh enzyme solution.

4. Using scissors or scalpel blades, mince the amniotic membrane to give pieces 1 cm^3 or less in size.

5. With the 50 ml volumetric pipette, transfer the tissue to a 500 ml Erlenmeyer flask containing a magnetic stirring bar. Add 50 ml of enzyme solution.

6. Incubate at 35°C with gentle mixing.

7. After 30 minutes remove the flask, allow the tissue to sediment, pipette off the supernatant and discard. Add 50 ml fresh enzyme solution and continue incubation at 35°C with gentle mixing.

8. After 30 minutes again separate the supernatant and place in a chilled centrifuge bottle. Store at 4°C.

9. Repeat the trypsinization cycle, continuing to harvest the supernatant fluid until the tissue is completely disaggregated or until you have harvested sufficient cells for your needs. Place each harvested aliquot in the refrigerator. All aliquots may be pooled if desirable.

10. When trypsinization is complete, centrifuge the collected samples at 500 RPM for 10 minutes. Discard the supernatant.

11. Resuspend the cells in 15-20 ml of complete growth medium ($Eagle_{80}$ Human $Serum_{20}$) and count the number of cells in an aliquot. Dilute with complete medium to a concentration of approximately $4x10^5$ cells/ml.

12. Plant roller tubes, Leighton tubes and French squares using 1 ml, 1 ml and 3 ml inocula respectively.

13. Incubate at 35°C and examine daily. Change medium as required.

NOTES

EXERCISE 3. CULTIVATION OF LEUCOCYTES FROM PERIPHERAL BLOOD

Monocytic leucocytes can be cultivated on a short term basis with little difficulty and serve as a readily available source of cells suited for a variety of experiments. The marked phagocytic activity of these cells has made them particularly useful for studies of host-parasite relationships employing intracellular bacterial agents such as *M. tuberculosis* (27) and *Br. abortus* (28).

Use of phytohemagglutinins to stimulate mitotic activity of leucocytes (29) has made feasible the use of leucocyte cultures for study of chromosomal abnormalities in somatic cells of humans as well as other animal species (30). The following procedure is based on the method recommended for human blood. It may be adapted to other species with appropriate modifications. Complete kits of materials for isolating leucocytes from whole blood, culturing them and using them for chromosomal analysis are available commercially.*

MATERIALS

1. One 10 ml sterile disposable hypodermic syringe with an #18 gauge, 3" needle.
2. Seventy percent ethanol, sterile cotton, tourniquet for bleeding.
3. Two ml of sterile heparin sodium solution containing 10 units/ml (100 μg/ml).
4. One half ml of sterile Bacto-Phytohemagglutinin.*
5. Four sterile 4 oz. French square bottles with rubber-lined screw caps.
6. Twenty-five ml of medium 199_{70} human $serum_{30}$ containing penicillin (50 units/ml) and streptomycin (50 μg/ml).
7. Four 5 ml and one 1 ml sterile, cotton plugged serological pipettes.

PROCEDURE

1. With a 1 ml pipette transfer 0.4 ml of phytohemagglutinin solution to the 2 ml of heparin. Mix well!

2. Draw 1.2 ml of heparin-phytohemagglutinin solution into the hypodermic syringe.

* Difco Laboratories, Detroit, Michigan.

NOTES

3. Draw 10 ml of venous human blood into the syringe containing heparin-phytohemagglutinin. Mix well! Replace the plastic cap over the hypodermic needle and stand the syringe in a vertical position with the needle up in a suitable holder. Allow to incubate at room temperature or at 35°C.

4. After 1-3 hours (or when at least 3 ml of leucocyte-plasma mixture is separated from the erythrocytes) remove the plastic needle guard and bend the hypodermic needle at a 90° angle. Holding the syringe in a vertical position carefully, push the plunger in and force the plasma-leucocyte layer into one of the 4 oz. French square bottles.

5. Count the number of cells in the plasma (p. 155) and dilute with growth medium to give a final concentration of $1.0 - 1.5 \times 10^6$ cells/ml.

6. Transfer 3-4 ml of cell suspension to each of three 4 oz. French square bottles and tighten the caps. The pH should be 7.4 at this time.

7. Incubate the cultures at 35°C. If the pH drops to 7.1 during incubation loosen the caps briefly to allow CO_2 to escape. The pH should be maintained between 7.1-7.4.

If the cells are to be used for chromosome analysis, colchicine should be added after 3 days incubation (final concentration of $0.5 - 1.0 \times 10^{-6}$ M). After 6 hours exposure to colchicine the cells can be loosened from the glass by vigorous agitation and treated as directed on page 199.

NOTES

EXPLANT CULTURES

Primary cultures may be initiated directly by explantation of source tissues without prior dispersion. The technique is particularly applicable when a limited amount of material is available (*e.g.,* tumor tissue, surgical biopsy) and is useful for the isolation of cell lines or for study of tissue tropisms, *etc.* This procedure obviates the need for dispersing tissues with enzymes or chelating agents and avoids the chemical and morphological alterations these materials may impose upon cells.

A number of techniques have been proposed for the cultivation of cells from explants. The need for a tightly closed vessel with satisfactory optical properties permitting microscopic examination led to the development of the Carrel flask and later to the Porter flask and the T-flask. An inexpensive flask, possessing many of the advantages of the above, is the "Agarslant"* culture tube. Gey (31) perfected the roller tube method in an attempt to culture larger quantities of tissue and to provide alternate bathing of the tissue and exposure to the atmosphere.

A third method of explant culture involves the use of a drop suspended from a coverglass which permits microscopic observation at high magnification. Such cultures are readily contaminated and are difficult to prepare. The perfusion technique (p. 60) is an extension of this method.

It is important to recognize that in any method of explant culture one must carefully select small (1-2 mm^3), uniformly cut explants. Use of explants of non-uniform size will make interpretation of results difficult and will preclude quantitative measurements of growth. Crushing or tearing of the edges of the explant will not only lead to non-uniformity but can interfere with migration and multiplication of the cells. Explants less than 1 mm^3 may contain too few cells to initiate growth while explants larger than 2 mm^3 tend to become necrotic as a result of inadequate diffusion of nutrients, gases and metabolic intermediates.

Proteolytic activity of the cells often causes liquefaction of the clot and may lead to a retraction of the cellular sheet. In such cases the clot must be "patched" (p. 44). The use of avian plasma is recommended to provide a firm clot even though it may be heterologous to the tissue under cultivation.

* Fischer Scientific Co., Pittsburgh, Pa.

NOTES

EXERCISE 1. FLASK AND TUBE TECHNIQUES

MATERIALS

1. One 9-11 day old chick embryo.
2. Five sterile "Agarslant"* culture tubes.
3. Five sterile #0 white rubber stoppers.
4. Fifteen sterile curved tip and fifteen straight capillary pipettes with rubber dropper bulbs.
5. One sterile 16 x 125 mm tube with rubber lined screw-cap.
6. Two sterile Bard-Parker handles with straight tipped blades (sharp).
7. Cockeral plasma (fresh or lypholized**) (p. 203).
8. Growth medium ($BSS_{40}Serum_{40}EE_{20}$) (p. 238).
9. Two sterile Petri dishes.
10. Dulbecco PBS (p. 217).
11. One rack for holding 16 mm tubes.

PROCEDURE

1. Aseptically remove tissue as rapidly as possible from the chick embryo. If other animals are used they preferably should be killed by some mechanical means.

2. Rinse tissue in PBS and place in a Petri dish in PBS containing 10% homologous serum.

3. Using two Bard-Parker handles with straight blades held scissor fashion, make rapid, clean cuts through the tissue. (Use a piece of paper ruled in mm squares to aid in judging size.) Be very critical and save only those explants which are regular in shape and with smooth cut edges.

4. When an adequate number of explants has been cut, wash them several times with PBS to remove debris. Pool the explants in medium ($BSS_{40}Serum_{40}EE_{20}$).

5. Using a bent tipped capillary pipette, spread 4-5 drops of plasma over the inner surface of each "Agarslant" tube. (Coat lower two-thirds of roller tubes with plasma.) Allow tubes to stand upright for a moment and draw off excess plasma.

* Fischer Scientific Co., Pittsburgh, Pa.

** Lypholized materials should be gassed with CO_2 to adjust pH upon reconstitution or pre-gassed reconstituting fluid used.

NOTES

6. With an unused pipette, pick up 5 or 6 explants in about 10 drops of medium and transfer to a tube. Thirty to forty explants arranged in rows can be used in the roller tube technique.

7. Mix the plasma and medium thoroughly and position the explants quickly. Tip the tube, withdraw and discard excess plasma and medium.

8. Allow culture vessel to lie horizontally until the plasma is coagulated. Add 1 ml of fluid medium and stopper tightly.

9. Incubate at 35°C and examine daily.

10. Proteolytic activity may cause the clot to liquify after a few days in which case it may be patched by:

    a. Removing used medium.

    b. Rinse explants with BSS.

    c. Spread a couple of drops of plasma over explants, clot by the addition of some EE_{50}.

    d. Add 1 ml of fresh medium.

11. When outgrowth has reached three or more times that of the original explant, the cultures should be transplanted:

    a. Cut explant and its outgrowth from clot with curved tip capillary pipette and lift out.

    b. Float colony on a small amount of medium in a Petri dish and allow to spread.

    c. Cut outgrowth by trimming edges (Cuts A, B, C, D on page 46).

    d. Cut outgrowth from original explant with a curved scalpel blade, using a rocking motion (Cuts E, F, G, H, on page 46).

    e. Cut outgrowth (1-8 page 46) into fragments 1 mm square. (Use only those fragments with smooth edges.)

    f. Replant fragments into fresh plasma as above in step 4.

NOTES

NOTE: The cultures may be transplanted by trypsinizing the culture and the cells used to initiate monolayer cultures. No specific formula can be given to guide one in feeding explant cultures. This is best done on a routine basis or as indicated by pH change.

A variation of this technique avoids the use of a plasma clot.

a. Proceed as above to cut explants (steps 1-3). Wash and pool in 3-4 drops of undiluted human cord serum.

b. Pick up explants and distribute them in tube. Allow tube to stand upright and draw off excess fluid.

c. Stopper loosely and allow to stand for approximately 30 minutes or until cord serum has dried down. Add medium as above and incubate in roller drum.

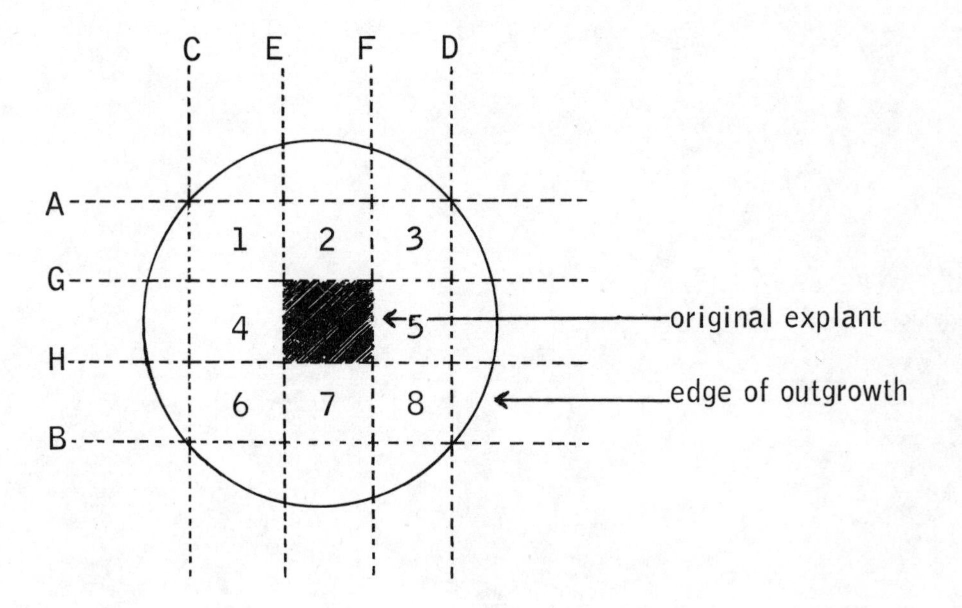

NOTES

EXERCISE 2. THE SLIDE CULTURE METHOD

Despite the distinct advantage of the slide culture method for microscopic observation, it should be recognized that the small volume of medium used is subject to dessication and concomitant changes in tonicity and pH.

MATERIALS

1. Five sterile Maximow slides.
2. Five sterile 44 x 60 mm #2 coverglasses.
3. Five sterile 22 mm round #1 coverglasses.
4. Fifteen sterile curved and fifteen sterile straight tipped capillary pipettes with rubber dropper bulbs.
5. Five sterile Petri dishes.
6. Sterile petrolatum (by dry heat sterilization).
7. Mixture of paraffin and petrolatum (3:1).
8. Camel hair brush (fine tip).
9. Two Bard-Parker handles with straight tipped blades.
10. Cockeral plasma (fresh or lypholized*) (p. 203).
11. Growth medium (BSS_{40} $Serum_{40}$ EE_{20}) (p. 238).
12. Dulbecco PBS (p. 217) containing 10% chicken serum.
13. Rack for 16 mm tubes.
14. One 9-11 day chick embryo.

PROCEDURE

1. Aseptically remove tissue (*e.g.,* lung, heart) as rapidly as possible. Rinse well with PBS containing 10% homologous serum.

2. Cut explants and wash as in steps 2 and 3, page 42.

3. Select suitable explants and pool in medium.

4. Place sterile 44 x 60 mm coverglass on a clean surface, add a drop of plasma to the coverglass and spread it over an area 20-30 mm in diameter using a curved tipped capillary pipette.

5. Pick up 3-4 explants in 4-5 drops of medium with a fresh curved tipped capillary pipette and place them on coverglass. Mix medium and plasma quickly (avoid bubbles in plasma); position explants and withdraw excess fluid.

* Lypholized materials should be gassed with CO_2 to adjust pH upon reconstitution or pre-gassed reconstituting fluid used.

NOTES

6. Allow the plasma to clot and then add 1-2 drops of fluid medium.

7. Touch a bit of sterile petrolatum to two corners of the coverglass (diagonally opposite). Invert a sterile Maximow slide over the culture centering the explants in the well. (Be sure to avoid the medium.)

8. Flip the slide over quickly to obtain a hanging drop. Seal carefully with the mixture of paraffin and petrolatum using a camel hair brush. Two or three coats may be required for a tight seal.

9. Incubate at 35° C and examine daily with the microscope.

10. Patch plasma clot as required.

NOTE: A modification of this procedure, involving a double coverslip, has an advantage in that the smaller coverslip (22 mm in diameter) bearing the culture may be removed and transplanted to a fresh slide or removed and stained at will. The procedure is as above except that the explants are placed on the 22 mm round coverslip, having spread plasma over its entire surface except for 2 mm around the edge. The round coverslip is then affixed, explants up, to a 44 x 60 mm coverglass with a droplet of water. Proceed as in step 5 on page 48.

NOTES

SPECIAL CELL CULTURE TECHNIQUES

EXERCISE 1. ISOLATION OF CLONES FROM ESTABLISHED CELL LINES

A clone is composed of a population of cells who are descendants of the same parent organism (32). The term can be refined by specifying that the parental organism must contain only a single nucleus. This qualification is not usually specified for mammalian cells although it may be implied. A variety of procedures can be employed for isolation of single cells. These include the techniques by which a single cell is aspirated into a capillary tube (33), or a micromanipulator is used to select a single cell, or a cell is dispensed in a microdrop (34) and allowed to propagate. All of these procedures assure that a single cell was isolated and the progeny derived therefrom was clonal.

Mammalian cells as monodisperse suspensions can be distributed in culture vessels, single cells visualized and marked, isolated by use of a suitable mechanical barrier, and allowed to grow to a population sufficiently dense to be trypsinized for serial propagation (34). The technique of plating cells to give clonal populations permits characterization of cell lines. The methodology is particularly well suited for quantitative studies because of its precision and amenability to statistical analysis. In this regard the term "plating efficiency" is an important parameter applied to a variety of clonal plating techniques. "Plating efficiency" refers to the percent of cells (viable) which form "clones" under a number of experimental conditions. Plating efficiencies can be scored in "absolute" terms or in "relative" values as per cent distributions. Absolute plating efficiency relates the total number of cells that grow to form clones to the number of cells contained in the original inoculum as determined by cell counts. Relative plating efficiency depends upon a comparison of the number of clones which develop under the imposed experimental conditions relative to controls. In both of these methods the procedures for tabulating data and recording dose-response relationships permit rigorous quantitation (32).

Maintenance of a clonal strain of cells requires selection for population homogeneity based on reliable genetic markers. Constant selection must be a routine procedure in which a rigorously standardized growth medium is employed. Markers that are useful include: cell morphology, colony morphology, chromosomal markers, nutritional requirements, growth rate, viral susceptibility, *etc.* (35).

NOTES

MATERIALS

A. 1. One monolayer culture of HeLa cells in $Eagle_{80}$ human $serum_{20}$ (p. 11). These cells should be in the mid-logarithmic phase of growth.
2. Ten ml of sterile 0.2% trypsin solution (p. 240, stock solution).
3. Forty five ml of sterile $Eagle_{80}$ human $serum_{20}$ (p. 237).
4. Ten sterile 2 oz. French square bottles with rubber lined screw caps.
5. Six sterile antibiotic assay cylinders* with one edge coated with silicone stopcock grease**.
6. Haemocytometer.
7. Five sterile screw-cap 16x125 mm culture tubes, each containing 4.5 ml of $Eagle_{80}$ human $serum_{20}$ (p. 237). These are dilution blanks.
8. Ten 1 ml, one 5 ml and four 10 ml sterile cotton plugged serological pipettes.

B. 1. Three bottles with isolated clones from A above.
2. Six sterile 2 oz. French square bottles with rubber lined screw caps.
3. Five ml of 0.05% trypsin (p. 240).
4. Eighteen sterile curved-tip capillary pipettes with rubber dropper bulbs.
5. Two sterile cotton plugged 5 ml serological pipettes.
6. Ten ml of $Eagle_{80}$ human $serum_{20}$.

PROCEDURE

A. 1. Trypsinize the monolayer of HeLa cells using 0.2% trypsin. Incubate at room temperature with occasional agitation until the cells just come off the glass.

2. With a 10 ml pipette triturate to yield a uniform dispersion of cells. Do both a total cell count (p. 156) and a viable count (erythrosin B dye exclusion technique, p. 157).

3. Using the dilution blanks prepare serial 10-fold dilutions (add 0.5 ml of cell suspension to 4.5 ml of medium in dilution blank) of the cell suspension until you have approximately $1x10^2$ viable cells per ml.

4. Transfer the 5 ml of suspension containing $1x10^2$ viable cells/ml to the bottle containing 45 ml of growth medium. This will constitute an additional 10-fold dilution and will yield a suspension containing approximately 10 viable cells per ml. Mix well.

* Stainless steel cylinder, OD 1 cm, ID 0.5 cm, 1 cm high.
** Dow-Corning.

NOTES

5. Pipette 4 ml of the cell suspension into each of 10 two oz. French square bottles. Tighten the caps and incubate at 35°C.

6. After 24 hrs. select three bottles at random and examine microscopically for single isolated cells. Mark a circle around two well isolated cells in each bottle. Aseptically place a sterile stainless steel antibiotic assay cylinder, siliconed side down, over each marked cell and press down firmly. Return the bottles to the incubator and continue incubation, undisturbed, for 14 days. The remaining 7 bottles should not be disturbed at this time and should likewise be incubated an additional 14 days.

B. 1. After 14 days incubation, harvest the clones isolated within the cylinders as follows:

With a capillary pipette remove the nutrient fluid from within the cylinder. Replace with 10 drops of 0.05% trypsin solution and incubate for 15 minutes at 35°C.

2. With a capillary pipette carefully take up and expel the trypsin solution 5-6 times. Finally take up the trypsin solution containing the cells and transfer, *in toto,* to a 2 oz. French square bottle. Add 3 ml of growth medium and stopper tightly. Incubate at 35°C for 14 days. Repeat for each of the six isolated clones.

3. Stain each of the 7 bottles from the original experiment (containing no cylinders) with Wright stain (p. 181).

4. Examine microscopically and list the following information in a suitable table.

   a. Number of clones per bottle.

   b. Mean absolute plating efficiency (based on number of viable cells planted).

   c. The per cent distribution of clones of the following types (36).

   Large compact

   Large diffuse

   Small compact

   Small diffuse

NOTES

5. After the appropriate 14 day incubation period repeat steps 3 and 4 for each of the six bottles prepared from isolated clones in step 1. Compare the per cent distribution of clonal variants in these "clonal" lines with that observed in the original 7 bottles representing the "wild" population. What conclusions can you draw from your results?

Higher plating efficiencies and more precise results may be obtained by substituting 60 mm petri dishes for the 2 oz. bottles used in this experiment. If this is done 5-6 ml of a cell suspension containing 10 cells/ml should be added to each dish and they should be incubated in a moist atmosphere containing 5% CO_2 in air.

NOTES

EXERCISE 2. PERFUSION CHAMBER TECHNIQUE

Time-lapse cinematography is an excellent method for studying and recording the normal activity of single cells *in vitro* when used in conjunction with a perfusion chamber. In addition, the effect of agents such as drugs, hormones, detergents or carcinogens may be demonstrated in the living cell with this technique.

A number of ingenious and practical designs for continuous perfusion have been proposed (37-39). Essentially all such systems consist of a chamber in which (a) cells may be maintained for short periods of time, (b) the fluid medium may be perfused over the cells in an interrupted or continuous flow from an external reservoir and (c) the cells may be visualized easily with phase contrast optics.

The Sykes-Moore perfusion chamber (40) is an especially convenient system. Depending upon the technique to be used to add cells, the chamber can be autoclaved completely disassembled or only partially assembled. In the method employed here, the cells will be grown on the upper coverglass in a widemouth French square bottle. When the monolayer is established, the coverglass is aseptically transferred in an inverted position to the partially assembled sterile chamber and the top screwed on to complete the assembly. The chamber is then filled with media. It should be pointed out there are two sizes of silicone rubber "O" ring gaskets for the chamber. The thicker of the two gives a 2. 5 mm working distance and is preferred for routine work. The thinner (1. 5 mm working distance) gasket is particularly recommended for use in time lapse cinematography where Köhler illumination is required.

MATERIALS

1. One Sykes-Moore perfusion chamber*.
2. Three sterile round coverglasses, 25 mm, #1*.
3. One "O" ring gasket, silicone rubber, 1 1/2 mm thick*.
4. One holder for perfusion chamber*.
5. One wrench assembly*.
6. One sterile curved forceps.
7. One petri dish, 60 mm.
8. One sterile 1 ml hypodermic syringe.
9. Two sterile hypodermic needles, 1", 23 gauge.
10. One sterile 4 oz. widemouth French square bottle with sterile #7 rubber stopper.
11. Monolayer culture of L-M cells in mid-log growth phase.
12. Twenty five ml sterile 199 P (p. 220).

* Bellco, Vineland, New Jersey.

NOTES

PROCEDURE

1. Assemble bottom half of perfusion chamber by placing a round coverglass inside of the deeper half of the chamber (part with inside threads). On top of the coverglass, place the "O" ring gasket. The assembled bottom assembly is placed inside a 60 mm petri dish. The top (outside threads) is placed alongside the bottom assembly with the wrench holes up. Sterilize by autoclaving.

2. Two round coverglasses are placed along one wall of a wide mouth 4 oz French square bottle and autoclaved with an aluminum foil cover.

3. Harvest cells as directed in exercise 1 (p. 14).

4. Add 10 ml of the suspension of L-M cells ($1.5\text{-}2 \times 10^5$ cells/ml) to the sterile French square bottle so the coverglasses are completely covered by the medium. Replace aluminum foil with sterile rubber stopper. Incubate at 35°C until the cells are attached to the coverglasses (a minimum of two hours). When the cells are attached, aseptically remove a coverglass with the sterile forceps, invert and place on top of sterile "O" ring gasket in the bottom assembly. The top part of the chamber is aseptically set in place and carefully tightened with a wrench.
   BE CAREFUL! DO NOT OVERTIGHTEN!

5. A sterile hypodermic needle is introduced into the chamber through one of the four access holes in the assembly and through the rubber gasket. This needle will provide an air escape when the chamber is filled.

6. Aseptically inject 1 ml of medium with the second needle introduced into the chamber as directed in step 5, using the hole opposite the air escape needle. The medium should be added SLOWLY in order to maintain cell attachment and to avoid creating bubbles within the chamber.

7. Remove both needles and place chamber in holder. Cells may be observed and photographed immediately or incubated at 35°C.

NOTES

EXERCISE 3. CELL AGGREGATION

Cells grown by conventional methods of explantation or by dispersed cell techniques tend to grow in patterns which have little gross resemblance to the histology seen *in vivo*. While the reasons for this type of growth are not completely understood, it is obvious that the physical and chemical conditions to which cells are exposed *in vitro* are quite different from those *in vivo*. Two important factors in the differentiation of cells are population size and density (41). When dissociated cells are cultured under prescribed conditions they will aggregate, interact and differentiate depending upon the potential of the cells used (42). Moscona (43) has devised a standardized technique for cell aggregation and has demonstrated that these dissociated cells will reassemble "histotypically."

MATERIALS

1. Sterile surgical instruments.
2. Five 1 ml and five 10 ml sterile, cotton plugged serological pipettes.
3. Two sterile Petri dishes.
4. Five sterile 10 ml screw-capped centrifuge tubes.
5. Two seven day embryonated hen's eggs.
6. Two sterile 25 ml Erlenmeyer flasks with rubber lined screw caps.
7. Thirty ml of culture medium. ($Eagle_{88}$ horse $serum_{10}$ $EE50_2$).
8. Ten ml of sterile 0.05% trypsin in calcium and magnesium free phosphated buffered saline (CMF-PBS).
9. One hundred ml of sterile CMF-PBS (p. 240).
10. Thirty ml of sterile Hanks BSS.
11. Two sterile 2 oz. French square bottles with rubber lined screw caps.

PROCEDURE

1. Candle 7-day embryonated eggs to determine viability. Remove embryos aseptically and transfer to sterile Petri dish. (steps 1-4, p. 106).

2. Wash embryos with sterile BSS to remove blood and yolk.

3. Remove organs to be cultivated (*e.g.*, retina, liver, limb bud devoid of skin or mesonephros) and transfer to a fresh Petri dish.

4. Mince tissue to obtain fragments approximately 0.5 - 1mm^3. Use scalpel blades in an opposed scissor-like motion.

5. Wash minced tissue twice with sterile BSS.

NOTES

6. Transfer minced tissue aseptically with a 1 ml pipette to a 10 ml screw-capped centrifuge tube containing five ml of CMF-PBS at pH 7.2 and incubate for 10 minutes at 35°C.

7. Gently centrifuge cells (500 RPM for 5 minutes) and decant off CMF-PBS.

8. Add five ml of 0.05% trypsin in CMF-PBS (pH 7.2) to centrifuge tube and incubate for 15-20 minutes at 35°C.

9. Again centrifuge gently and decant off trypsin solution.

10. Rinse three times with CMF-PBS by repeated gentle centrifugation and decanting. Be sure to avoid disruption of fragments at this stage.

11. Add 1 ml of culture medium to centrifuge tube and disperse fragments by triturating 10 to 15 times being careful not to allow the medium to froth.

12. Remove a 0.1 ml sample of cells. Count and check viability. (p. 155-156). Add fresh medium to give a final concentration of $5x10^5$ cells/ml.

13. Transfer 3 ml of the cell suspension to each of two 25 ml screw-capped Erlenmeyer flasks. Gas with 5% CO_2 in air and tighten caps. Incubate at 35°C on a rotary shaker at 70 RPM. (One inch radial stroke).

14. Transfer 3 ml of the cell suspension (from 12 above) to each of two 2 oz. screw-capped French squares. Tighten the caps and incubate in a horizontal position at 35°C with the most uniform surface down. Examine after 7 days.

15. Harvest aggregates from suspension cultures (step 13) after 24 and 48 hours. Count the number of aggregates and measure their size. Fix in Bouin fluid, section, stain and examine.

16. Aggregates in bottles (step 14) may be excised and subsequently fixed and sectioned or the entire bottle may be fixed and the cells stripped and stained by the collodion method (p. 179).

NOTES

REFERENCES

1. EVANS, V. J., and EARLE, W. R., The use of perforated cellophane for the growth of cells in tissue culture, J. Nat. Cancer Inst., 8: 103-119 (1947).
2. EARLE, W. R., SCHILLING, E. L., et al., Production of malignancy in vitro. X. Continued description of cells at the glass interface of the cultures, J. Nat. Cancer Inst., 10: 1067-1103 (1950).
3. SANFORD, K. K., EARLE, W. R., et al., The measurement of proliferation in tissue cultures by enumeration of cell nuclei, J. Nat. Cancer Inst., 11: 773-795 (1951).
4. EVANS, V. J., EARLE, W. R., et al., The preparation and handling of replicate tissue cultures for quantitative studies, J. Nat. Cancer Inst., 11: 907-927 (1951).
5. EARLE, W. R., SANFORD, K. K., et al., The influence of inoculum size on proliferation in tissue cultures, J. Nat. Cancer Inst., 12: 133-153 (1951).
6. HSU., T. C., and MERCHANT, D. J., Mammalian chromosomes in vitro. XIV. Genotypic replacement in cell populations, J. Nat. Cancer Inst., 26: 1075-1083 (1961).
7. GEY, G. O., COFFMAN, W. D., et al., Tissue culture studies of the proliferative capacity of cervical carcinoma and normal epithelium. Cancer Res., 12: 264-265 (1952) (abstract).
8. MONOD, J., The growth of bacterial cultures. Ann. Rev. Microbiol., 3: 371-394 (1949).
9. OWENS, O. von H., GEY, M. K., et al., Growth of cells in agitated fluid medium, Ann. New York Acad. Sc., 58: 1039-1055 (1954)
10. EARLE, W. R., SCHILLING, E. L., et al., The growth of pure strain L cells in fluid-suspension cultures, J. Nat. Cancer Inst., 14: 1159-1171 (1954).
11. McLIMANS, W. F., GIARDINELLO, F. E., et al., Submerged culture of mammalian cells: the five liter fermentor, J. Bact., 74: 768-774 (1957).
12. ZIEGLER, D. W., DAVIS, E. V., et al., The propagation of mammalian cells in a 20-liter stainless steel fermentor, Appl. Microbiol., 6: 305-310 (1958).
13. RIGHTSEL, W. A., McCALPIN, H., et al., Studies on large-scale methods for propagation of animal cells, J. Biochem. & Microbiol. Technol. & Engin., 2: 313-325 (1960).
14. KUCHLER, R. J., and MERCHANT, D. J., Propagation of strain L (Earle) cells in agitated fluid suspension cultures, Proc. Soc. Exper. Biol. & Med., 92: 803-806 (1956).
15. LEWIS, M. R., and LEWIS, W. H., The growth of embryonic chick tissues in artificial media, agar and bouillon, Bull. Johns Hopkins Hosp., 22: 126-127 (1911).
16. WEIL, G. C., Some observations on the cultivation of tissues in vitro, J. Med. Res., 26: 159-180 (1912).
17. ZINSSER, H., WEI, H., et al., Agar slant tissue cultures of typhus rickettsiae (both types), Proc. Soc. Exper. Biol & Med., 37: 604-606 (1937).
18. CHEEVER, F. S., Cultivation of Herpes febrilis virus on agar slant tissue cultures, Proc. Soc. Exper. Biol. & Med., 42: 113-114 (1939).
19. WALLACE, J. H., and HANKS, J. H., Agar substrates for study of microepidemiology and physiology in cells in vitro, Science, 128: 658-659 (1958).
20. ZITCER, E. M., FOGH, J., et al., Human amnion cells for large-scale production of polio virus, Science, 122: 30 (1955).
21. WEINSTEIN, H. J., ALEXANDER, C., et al., Preparation of human amnion tissue cultures, Proc. Soc. Exper. Biol. & Med., 92: 535-538 (1956).
22. TAKEMOTO, K. K., and LERNER, A. M., Human amnion cell cultures; susceptibility to viruses and use in primary virus isolations, Proc. Soc. Exper. Biol. & Med., 94: 179-182 (1957).
23. BECKER, Y., GROSSOWICZ, N., et al., Metabolism of human amnion cell cultures infected with poliomyelitis virus. I. Glucose metabolism during virus synthesis. Proc. Soc. Exper. Biol. & Med., 97: 77-82 (1958).
24. FERNANDES, M. V., The development of a human amnion strain of cells, Texas Rep. Biol. & Med., 16: 48-58 (1958).
25. YOUNGNER, J. S., Monolayer tissue cultures. I. Preparation and standardization of suspensions of trypsin-dispersed monkey kidney cells, Proc. Soc. Exper. Biol: & Med., 85: 202-205 (1954)

26. HAYASHI, H., and LoGRIPPO, G. A., Preparation of primary human amnion cells by pancreatin procedure and their susceptibility to enteroviruses, J. Immunol., 90: 956-959 (1963).

27. SUTER, E., The multiplication of tubercle bacilli within normal phagocytes in tissue culture, J. Exper. Med., 96: 137-150 (1952).

28. POMALES-LEBRÓN, A., and STINEBRING, W. R., Intracellular multiplication of Brucella abortus in normal and immune mononuclear phagocytes, Proc. Soc. Exper. Biol. & Med. 94: 78-83 (1957).

29. HUNGERFORD, D. A., DONNELLY, A. J., et al., The chromosomes constitution of a human phenotypic intersex, Am. J. Human Genet., 11: 215-236 (1959).

30. MOORHEAD, P. S., NOWELL, P. C., et al., Chromsome preparations of leukocytes cultured from human peripheral blood, Exper. Cell Res., 20: 613-616 (1960).

31. GEY, G. O., An improved technic for massive tissue culture, Am. J. Cancer, 17: 752-756 (1933).

32. PUCK, T. T., MARCUS, P. I., et al., Clonal growth of mammalian cells in vitro: Growth characteristics of colonies from single HeLa cells with and without a "feeder" layer, J. Exper. Med., 103: 273-284 (1956).

33. SANFORD, K. K., EARLE, W. R., et al., The growth in vitro of single isolated tissue cells, J. Nat. Cancer Inst., 9: 229-246 (1948).

34. LWOFF, A., DULBECCO, R., et al., Kinetics of the release of poliomyelitis virus from single cells, Viology, 1: 128-139 (1955).

35. MERCHANT, D. J., and NEEL, J. V., (eds.) Approaches to the Genetic Analysis of Mammalian Cells, Univ. Michigan Press (1962).

36. MURPHY, W. H., and LANDAU, B. J., Clonal variation and interaction of cells with viruses, Nat. Cancer Inst. Monogr., 7: 249-271 (1962).

37. PAUL, J., A perfusion chamber for cinemicrographic studies, Quart. J. Microscop. Sc., 98: 279-280 (1957).

38. POMERAT, C. M., Perfusion chamber, Meth. M. Res., 4: 275-277 (1951).

39. ROSE, G. G., Special uses of the multipurpose tissue culture chamber, Texas Rep. Biol. & Med., 15: 310-312 (1957).

40. SYKES, J. A. and MOORE, E. B., A simple tissue culture chamber, Texas Rep. Biol. & Med., 18: 288-297 (1960).

41. GROBSTEIN, C., Differentiation of vertebrate cells, in Brachet, J. and Mirsky, A. E., (eds.) The Cell, New York, Academic Press, Vol. 1, pp. 437-496 (1959).

42. MOSCONA, A., The development in vitro of chimeric aggregates of dissociated embryonic chick and mouse cells, Proc. Nat. Acad. Sc., 43: 184-193 (1957).

43. MOSCONA, A., Rotation-mediated histogenetic aggregation of dissociated cells: a quantifiable approach to cell interactions in vitro, Exper. Cell Res., 22: 455-475 (1961).

Chapter III.
ORGAN CULTURE METHODS

In a further attempt to bridge the gap between cell culture and the *in vivo* condition, the method of organ culture is being increasingly utilized. This technique involves the cultivation of fragments of organs or entire embryonic organs in a tissue culture system. Ultimately, one endeavors to deal with normal tissue relationships as they exist in the body, divorced from the complexities of organ interaction. The problem resolves itself into the maintenance of differentiated cells as a group of normally associated tissues. The technique is designed to inhibit outgrowth of cells from the explant in contrast to standard cell culture procedures.

Historically, the technique for growth and differentiation of organs *in vitro* was developed by T. S. P. Strangeways and Honor B. Fell (1) to study bone development. Until recently, the method has been utilized extensively as a purely morphological tool (2-4). In the last few years, the technique has been adopted for physiologic studies involving vitamins (5-7), hormones (7-12), carcinogens (13) and embryonic inductors (14). Many aspects of this technique have been reviewed recently (15, 16).

Explants for organ culture should be obtained aseptically and as rapidly as possible from the animal, preferably killed by some mechanical means. In general, organs from younger animals are more easily cultivated. Since the maximum size that may be maintained successfully is of the order of 2 mm^3, it is necessary with larger organs to use fragments. Considerable care should be taken in dissecting such explants to avoid tearing or crushing of the tissues. A convenient method of cutting is to use two scalpels in a scissor-like fashion. Sharp instruments are essential! In addition, recognition should be given to the normal architecture and histogenesis of the organ being cultivated. For example, in cutting fragments of organs that have histogenic polarity in their normal differentiation (*e. g.,* skin, ovary, *etc.*), it is essential to provide each explant with a normal complement of tissue types.

Two exercises are provided as representative of the many methods available (4, 15-17). Explants must be subcultured periodically (2 days for chick tissues, 3 days for mammalian) and all explants should be checked daily, noting changes in density, outgrowth and size. Camera lucida drawings should be made if gross differentiation of the explants is being followed. Periodically, explants should be fixed and stained for general histology (p. 176) or other histochemical approaches may be used (p. 188). The media may be saved and used for chemical studies.

NOTES

EXERCISE 1. THE PLASMA CLOT TECHNIQUE

The proportion of plasma and embryo extract used will vary in this technique with the tissue under study. In general, one uses a lower proportion of EE than in other types of culture. In addition, embryo extract from a 12-14 day chick is recommended since extracts prepared from embryos of a younger age tend to stimulate cellular outgrowth at the expense of differentiation.

MATERIALS

1. Two sterile plastic organ culture dishes. *
2. Two sterile fine forceps (Iris or jewelers).
3. Two sterile scalpels (Iris, or Bard-Parker #10).
4. Dozen sterile cotton plugged capillary pipettes.
5. Five sterile 10 ml serological pipettes.
6. Cockeral plasma (p. 203). (See note bottom of p. 42).
7. Embryo extract (p. 205).
8. Chick embryo, 6-8 days incubation.
9. Two sterile 100 mm Petri dishes
10. Twenty-five ml sterile BSS (p. 216).

PROCEDURE

1. Add 5 ml of sterile distilled water to absorbent disc to provide a humid atmosphere. Do not allow water to flow into the center well.

2. Add approximately 9 drops of plasma to the center well followed by 3 drops of EE_{50} and quickly but gently mix before medium clots. Be careful not to create bubbles.

3. Allow the mixture to clot.

4. Aseptically remove embryo (p. 205) from its shell and transfer to a sterile Petri dish. Dissect out tissue to be cultivated (*e. g.*, urogenital tract, auditory vesicles, skin). If necessary, cut explants to approximately 2 mm^3. Rinse in BSS.

5. Carefully pipette fragments onto clot (usually 4 explants per dish). Remove the excess BSS carried over using a fine pipette.

* Falcon Plastics, Division of B-D Laboratories, Inc., Los Angeles, California.

NOTES

6. Incubate at 35° C. (Tissues may vary in their temperature requirements.) A strip of thin wood placed below the organ culture dish in the incubator will prevent condensation of water on the inner surface of the dish cover.

7. Subculture every second day by carefully cutting clot and any outgrowth away from the explant. Wash in two or three changes of BSS and transfer to a freshly prepared plasma clot.

8. After 6-7 days, carefully cut explant away from clot and fix in Bouin fluid (p. 178). Tissues should be sectioned and stained (p. 185).

NOTES

EXERCISE 2. THE RAFT TECHNIQUE

The use of fluid medium is possible in organ culture if adequate support, such as a raft, is provided for the explant and the explant is positioned at the gas-medium interface. A number of methods have been proposed (15, 16); the use of a stainless steel mesh is presented here as representative. This technique makes possible the use of defined media and thus permits characterization of the action of various agents, *viz.*, vitamins, hormones, viruses, *etc.* Although a number of media have been used for this technique, a modified Trowell medium T8 (p. 228) is recommended for this exercise (17). Biggers (18) medium BGJb or Waymouth's 752/1 (p. 229) may be substituted.

MATERIALS

1. Two sterile plastic organ culture dishes.*
2. Two sterile fine forceps (Iris or jewelers).
3. Two sterile scalpels (Iris or Bard-Parker #10).
4. Three sterile, cotton plugged capillary pipettes.
5. Two sterile rafts of stainless steel wire mesh.*
6. One sterile 10 ml and one 1 ml serological pipettes.
7. Trowell medium T8 (p. 228).
8. Three week old mouse.
9. Twenty-five ml sterile BSS (p. 216).
10. Two sterile 100 mm Petri dishes.

PROCEDURE

1. Moisten absorbent disc in culture vessels with 5 ml sterile distilled water.

2. Aseptically remove organ to be cultivated (*e.g.*, lymph node, ovary, thyroid) and cut into explants 2 mm^3. Wash in sterile BSS.

3. Pipette explants (usually 4) aseptically onto a stainless steel raft in a sterile Petri dish using as small a droplet of BSS as possible. Remove any excess fluid. Pick up raft with sterile forceps and place carefully over center well of an organ culture dish.

4. Add approximately 0.9 ml of medium to center well. Be sure that the raft and tissue are in contact with the medium but not submerged.

* Falcon Plastics, Division of B-D Laboratories, Inc. Los Angeles, California.

NOTES

5. Incubate at 35 C. Gas with 5% CO preferably in 95% oxygen (17) to maintain correct pH and provide sufficient oxygenation of the entire explant. Embryonic and neoplastic tissue (17) however, are adversely affected by high oxygen tensions.

6. Subculture every third day by aspirating medium from below raft. (This may be kept for assay.) Replace with fresh medium.

7. After 6-7 days, fix explants (p. 178). Tissues must be sectioned and stained (p. 185).

NOTES

REFERENCES

1. STRANGEWAYS, T. S. P., and FELL, H. B., Experimental studies on the differentiation of embryonic tissues growing in vivo and in vitro. I. The development of the undifferentiated limb bud (a) when subcutaneously grafted into the post embryonic chick and (b) when cultivated in vitro., Proc. Roy. Soc., London, Ser. B., 99: 340-366 (1926).
2. FELL, H. B., Recent advances in organ culture, Sc. Progr., 41: 212-231 (1953).
3. GAILLARD, P. J., Growth and differentiation of explanted tissues, Internat. Rev. Cytol., 2: 331-401 (1953).
4. MARTINOVITCH, P. N., A modification of the watch glass technique for the cultivation of endocrine glands of infantile rats, Exper. Cell Res., 4: 490-493 (1953).
5. KAHN, R. H., Vaginal keratinization in vitro. Ann. New York Acad. Sc., 83: 347-355 (1959).
6. LASNITZKI, I., Effect of excess vitamin A on the normal and oestrone treated mouse vagina grown in chemically defined medium, Exper. Cell Res., 24: 37-45 (1961).
7. FELL, H. B., DINGLE, J. T., et al., Studies on the mode of action of excess of vitamin A.: 4. The specificity of the effect of embryonic chick-limb cartilage in culture and on isolated rat liver lysosomes. Biochem. J., 83: 63-69 (1962).
8. MEITES, J., KAHN, R. H., et al., Prolactin production by rat pituitary in vitro, Proc. Soc. Exper. Biol. & Proc., 108: 440-443 (1961).
9. RIVERA, E. M., and BERN, H. A., Influence of insulin on maintenance and secretory stimulation of mouse mammary tissues by hormones in organ-culture, Endocrinology, 69: 340-353 (1961).
10. ELIAS, J. J., Response of mouse mammary duct end-buds to insulin in organ culture, Exper. Cell Res., 27: 601-604 (1962).
11. FELL, H. B., The influence of hydrocortisone on the metaplastic action of vitamin A on the epidermis of embryonic chicken skin in organ culture, J. Embryol. & Exper. Morphol., 10: 389-409 (1962).
12. HAGMÜLLER, K. and LESLIE, I., The use of organ culture to study ^{131}I uptake and metabolism of thyroid tissue, Exper. Cell Res., 27: 396-400 (1962).
13. LASNITZKI, I., Organ culture as a tool in the study of carcinogenesis, Acta Unio. Internat. Contra. Cancrum, 15: 627-631 (1959).
14. GROBSTEIN, C., Interactive processes in cytodifferentiation, J. Cell & Comp. Physiol. 60 (Suppl. 1): 35-48 (1962).
15. WOLFF, M. E. (ed), La Culture Organotypique; Paris, Centre National de la Recherche Scientifique, (1961).
16. DAWE, C. J., (ed), Symposium on Organ Culture: Studies of Development, Function and Disease, Nat. Cancer Instit. Monogr., 11: 1-252 (1963).
17. TROWELL, O. A., The culture of mature organs in a synthetic medium, Exper. Cell Res., 16: 118-147 (1959).
18. BIGGERS, J. D., GWATKIN, R. B. L., et al., Growth of embryonic avian and mammalian tibiae on a relatively simple chemically defined medium, Exper. Cell. Res., 25: 41-58 (1961).

Chapter IV.
APPLICATIONS OF CELL AND ORGAN CULTURE

VIROLOGY

EXERCISE 1. CYTOPATHIC EFFECTS OF VIRUSES

The cytopathic effects (CPE) of viruses are often sufficiently distinctive and regular in occurrence to provide a means for distinguishing various groups of viruses or for characterizing cell lines (1, 2). Changes in cellular morphology caused by adverse nutritional and/or environmental abnormalities must not be confused with viral CPE. The cardinal characteristics of viral induced cytopathic changes may be summarized as follows:

Rapidity and nature of cell destruction: The polioviruses are markedly cytopathic (necrotic) for sensitive cells; a single virus particle can destroy an infected cell in 8 to 12 hours. On the other hand, viruses which cause intranuclear inclusions are less cytopathic than the polioviruses, *e.g.*, herpes simplex virus or the adenoviruses. Some viruses, such as that of measles, take a prolonged period of time to cause destruction of cell cultures. Such agents initially manifest their CPE by inducing focal areas of degeneration characterized by coalescence of contiguous cells (syncitia); eventually the degenerative changes encompass the entire culture. Some viruses can infect cell cultures without causing observable morphologic changes and indirect methods, such as the hemadsorption test, must be used to detect viral infection.

Inclusion bodies: Structures within cells consisting of aggregates of virus particles, or structures which result as a cellular response to viral infection, are called "inclusion bodies". The presence or absence of inclusion bodies in cells is an important parameter of virus-cell interaction. By use of polychrome dyes it is possible to allocate viral inclusion bodies into two groups, *viz.*, those which are basophilic and those that selectively stain with acid dyes (acidophilic). Basophilic inclusions occur usually in the cytoplasm of cells; acidophilic inclusion bodies may be found in the nucleus or cytoplasm depending upon the agent in question.

MATERIALS

1. Eight Leighton tubes, each containing a monolayer of HeLa cells grown on coverslips (p. 14). A strain of HeLa cells adapted to growth in Eagle medium 80% calf serum 20% (p. 237) should be used; each tube should contain 0. 9 ml of medium.

2. Two-tenths ml (200 TCID, *i. e.*, tissue culture infectious doses) of each of the following viruses; vaccinia, herpes simplex, and polio-virus, type 2*.

3. Trays for sterilization of supernatant fluids from virus-infected cultures.

PROCEDURE

1. Label two tubes vaccinia, two polio, two herpes and two controls.

2. Inoculate 0. 1 ml of the appropriate virus into each pair of tubes.

3. Incubate tubes at 35° C until definite cytopathic changes ensue; do not permit destruction to become too extensive.

4. Decant supernatant fluids from tubes, aseptically remove the cover-slips, and stain them by the hematoxylin-eosin procedure (p. 185).

5. Examine stained preparations for inclusion bodies and make representative drawings.

NOTE: The student is advised to use adequate precautions to prevent contamination of his person or environs with infectious material.

* American Type Culture Collection, Washington, D. C.

NOTES

EXERCISE 2. ISOLATION OF VIRUSES FROM CLINICAL SPECIMENS: IDENTIFICATION BY NEUTRALIZATION TESTS

The purpose of this exercise is to demonstrate that viruses can be isolated by cell culture techniques (3) from heavily contaminated clinical specimens. Because the subject of this exercise encompasses diagnostic virology generally, it is necessary to use a representative example to illustrate effective procedures, *e. g.*, isolation of an enteric virus from a fecal specimen and identification by neutralization tests.

The choice of cells to be used for isolation of virus is a function of the host-range of the virus under study. For example, cell cultures prepared from monkey kidney, human amnion, or established cell lines, such as HeLa cells, can be employed for the polioviruses. One must remember that a single specimen may yield no virus at all or one or more unrelated viruses. Identification of virus is accomplished by the selective use of homotypic specific antiserum which prevents destruction of cells by virus.

Grossly contaminated specimens can be made suitable for inoculation into cell cultures by simple differential centrifugation providing that antibiotics are included in the suspending fluid. Because filters adsorb significant quantities of virus and could give rise to false negative results, they generally are not used for diagnostic work. The ordinary clinical centrifuges are satisfactory for diagnostic procedures but care must be exercised in removal of supernatant fluids to prevent resuspension of sedimented materials. With heavily contaminated specimens it may be necessary to remove gross debris by a preliminary centrifugation.

Because antiserum prevents infective attachment of virus to cell receptors (neutralization of viral infectivity) neutralization techniques have had two common applications, *viz.*, use of known antiserum to identify unknown virus or measurement of the concentration of antibody in sera to known viruses. In either case one mixes antiserum and virus under standardized conditions and assays the reaction mixture for free (infective) virus. Because sera may contain antibodies which cause nonspecific virus neutralization, block cell receptors, or are cytotoxic, careful work requires either a variety of essential controls or preliminary tests. The theory of neutralization tests employed to titrate the antibody content of sera is complex (4, 5, 6).

MATERIALS

1. Nine 16 x 125 mm tubes with rubber lined screw caps containing 24 to 48 hr. monolayer cultures of HeLa cells grown in $Eagle_{80}$ calf $serum_{20}$ (p. 237).

NOTES

2. Fecal specimen (10% suspension in BSS containing, in final concentration, penicillin 100 units and streptomycin 100 μg/ml) thought to contain poliovirus, type 2.

3. Three-tenths ml of sterile antiserum to poliovirus, type 2 (7).

4. Three 1 ml and one 10 ml sterile cotton plugged serologic pipettes.

5. Culture tube rack.

6. Tray for contaminated materials.

7. One sterile 15 ml screw-cap conical centrifuge tube.

PROCEDURE

NOTE: The student is advised to use adequate precautions to prevent contamination of his person or environs with infectious material.

1. Label tubes 1-3 cell controls, tubes 4-6 virus, and tubes 7-9 virus plus antiserum.

2. Centrifuge the fecal specimen (10 ml) in a clinical centrifuge at maximum speed for 30 minutes.

3. After centrifugation very carefully remove 0.6 ml of supernatant fluid. (Do not touch the sides of the tube with the pipette or disturb the sedimented materials.)

4. Add 0.1 ml of the supernate to tubes 4-9.

5. Add 0.1 ml of antiserum to tubes 7-9.

6. Incubate the tubes at 35° C and inspect them daily for 7 days. Note which cells undergo progressive cytopathic changes compared with the controls. Tabulate the data in a suitable table.

NOTES

EXERCISE 3. TITRATION OF MAMMALIAN VIRUSES BY THE TUBE METHOD

Tissue culture techniques have expanded the scope of virology by: (a) providing susceptible host systems for detection and assay of viruses that were previously unknown; (b) affording, in quantity, uniformly susceptible cell populations as a substitute for tedious, imprecise and expensive animal bioassay procedures; (c) isolation of virus-cell interaction so that biochemical and genetic studies can be conducted without the complicating influences contributed by host reaction to infection.

A variety of procedures is available for titration of viruses by application of cell culture techniques; the precision of assay and the problem under investigation dictate which procedure is most applicable. When very precise measurements of virus concentration are not required and large numbers of samples are to be assayed, the "metabolic inhibition test" employing plastic panels has particular advantages (8). Cells and dilutions of virus are dispensed into small cups in plastic panels. Cells in cups that do not contain virus continue to metabolize; acid production causes a change in the phenol red indicator from its red color at pH 7.4 to yellow; similar pH change does not occur in cups containing virus infected cells.

Tube titrations are essentially the same as panel titrations. As a consequence of viral infection, cells undergo characteristic morphologic and metabolic changes. Endpoints in tube and panel titrations are scored according to the cytopathic effect of virus and/or accompanying pH changes. Virus concentration (infectious units per ml of undiluted virus) usually is calculated according to the method of Reed and Muench or the "most probable number" method (4, 6). Technical problems associated with titrations of virus are simplified appreciably by use of a growth medium which contains no serum or serum free of antibody to the virus under study, in which case no special maintenance media are required (9).

MATERIALS

1. Seventeen culture tubes of HeLa cells grown in 0.9 ml of $Eagle_{80}$ calf $serum_{20}$ (p. 237).

2. Five-tenths ml of each of three ten-fold dilutions of poliovirus, type 2; the dilution range should be such that the highest concentration of virus will cause uniform cell lysis and the lowest no lysis. For example, if dilutions 10^{-4}, 10^{-5}, 10^{-6} of virus are employed the lowest dilution (10^{-4}) should lyse the cells in all of the tubes.

3. Five sterile cotton plugged 1 ml serologic pipettes.

NOTES

4. Container for contaminated pipettes.

PROCEDURE

1. Inoculate 0.1 ml of each virus dilution into each of five tubes; save two tubes as uninoculated cell controls. BE CERTAIN THAT TUBES ARE STOPPERED TIGHTLY BEFORE INCUBATION AT 35° C.

2. Read tubes daily for 7 days and score results according to progressive cytopathic changes; suggestive virus induced morphologic changes (1 +); definite morphologic changes (2 +); almost complete cellular degeneration (3 +); complete cellular destruction (4 +).

NOTE: The student is advised to use adequate precautions to prevent contamination of his person or environs with infectious material.

NOTES

EXERCISE 4. TITRATION OF MAMMALIAN VIRUSES BY THE PLAQUE METHOD

To titrate viruses by the plaque technique, advantage is taken of the fact that a virus particle (plaque forming unit or PFU) can infect a single cell, multiply, and spread to contiguous cells. The localization of these necrotic cells in a monolayer gives rise to a circumscribed lesion known as a plaque. A solid (agar) or semi-solid (methylcellulose) overlay medium is used to prevent diffusion of virus particles. The plaque technique (10) for titration of viruses has been adopted because it is precise, permits isolation of clones of virus possessing distinctive biological properties, and is more amenable to statistical and genetic analyses of virus-cell interaction than are other assay methods.

Factors associated with the successful application of the plaque technique to titrations of mammalian viruses can be delineated as follows:

a. Hardiness of the indicator cell line will determine whether the cells can survive in maintenance medium for a period of time sufficient to permit lysis of cells by virus. It will also determine whether the monolayer can be maintained intact upon addition of the overlay;

NOTE: A modification of the agar method of plaque titration utilizes a suspension of tissue cells in agar rather than an agar overlay (11). This makes possible the use of a wider range of tissue cell lines for virus assay.

b. Inherent cytopathogenicity of the virus will govern whether the plaque titration is applicable;

c. Young, log phase cells at their physiologic optimum should be used. The cells comprising the monolayer should not be too tightly crowded together;

d. pH control is critical but can be managed by the use of appropriate buffer systems. It is expedient for long term or large scale operations to employ an incubator which provides a humidified atmosphere of 95% air and 5% CO_2;

e. The nutritional adequacy of the maintenance medium must be considered since susceptibility of mammalian cells to viral infection is influenced by their physiologic status;

f. The possible presence of toxic materials in agar must be recognized and appropriate measures taken to eliminate them. These problems can be avoided, to a major extent, by the use of a purified agar such

NOTES

as Noble agar. * It has been reported (12) that one important inhibitor is an acidic polysaccharide which can be neutralized by protamine sulfate.

MONOLAYER-AGAR OVERLAY

MATERIALS

1. Water bath adjusted to 48° C.

2. Ten ml of liquid agar overlay medium at a temperature of 45° to 50° C. The composition of the overlay medium is: 1 volume of melted, sterile 3% Noble agar in BSS: 3 volumes of Scherer maintenance medium 80%, calf serum 20%.

3. Three 2 oz. French square bottles, each containing young (4-5 day) monolayer cultures of HeLa cells (grown in $Eagle_{80}$ Human $Serum_{20}$) which have been washed 3 times with BSS.

4. Two 1 ml and two 10 ml sterile, cotton plugged, serological pipettes.

5. One ml of poliovirus, type 2, diluted in BSS containing 5% calf serum to give approximately 40 plaque forming units (PFU) of virus per ml.

6. Wright stain (p. 181).

PROCEDURE (13)

1. Add 0. 5 ml of virus suspension to each of two bottles of washed HeLa cells. The uninoculated bottle serves as the cell control. Incubate the bottles for 1 hour at 35° C.

2. Add 3 ml of the overlay medium to each bottle with a warm pipette (50° C). Spread the overlay medium evenly over the cells.

3. Incubate the cultures for 3 to 4 days at 35° C or until plaques are detectable by gross inspection.

4. Pour off the soft agar overlay into a container for autoclaving. If necessary the bottles may be agitated gently to loosen the agar layer.

* Difco Laboratories, Detroit, Michigan.

NOTES

5. Stain the exposed monolayer with Wright stain (p. 181) and examine the plaques microscopically using a 10X objective.

6. Count and record the number of plaques. How do the figures compare with the estimated amount of virus put into the bottles?

MONOLAYER-METHYLCELLULOSE OVERLAY

With certain viruses such as herpes simplex the presence of inhibitors in agar prevents maximum development of plaques. Even the addition of protamine sulfate fails to yield maximum plaque formation. A substitute overlay medium composed of methylcellulose in a nutrient solution may give up to a 10-fold increase in sensitivity of assay (12).

MATERIALS

1. Ten ml of sterile methylcellulose overlay medium. The composition of this overlay medium is:

   a. One volume of 3% methylcellulose* in distilled water. The methylcellulose is dissolved by autoclaving 3 grams of the powder in 100 ml of water at 18 lbs/in^2 for 20 minutes. After autoclaving the material is allowed to cool to 5° C and final solution is achieved by stirring with a magnetic bar.

   b. One volume of Eagle basal medium in Hanks BSS (p. 226).

   The Eagle medium and methylcellulose are stirred together to give a homogenous mixture.

2. Three 2 oz French square bottles, each containing young (4-5 day) monolayer cultures of HeLa cells (grown in Eagle$_{80}$ Calf serum$_{20}$) which have been washed 3 times with BSS.

3. Two 1 ml and two 10 ml sterile cotton plugged serological pipettes.

4. One ml of herpes simplex virus diluted in BSS containing 5% calf serum to give approximately 40 plaque forming units (PFU).

5. Twenty ml of sterile BSS.

6. Wright stain (p. 181).

* Dow Chemical Co., 4000 CPS, USP.

NOTES

PROCEDURE

1. Add 0. 5 ml of virus to each of two bottles of washed HeLa cells. The uninoculated bottle serves as the cell control. Incubate the bottles at 35° C for 1 hour to permit virus attachment.

2. Add 3 ml of overlay medium and distribute evenly over the surface of the monolayer.

3. Incubate the cultures at 35° C for 5-6 days.

4. After 5-6 days incubation add 3 ml of BSS to each bottle. Gently rock the bottles to mix the BSS and methylcellulose. This will dilute the overlay so that it can be decanted readily.

5. Decant the diluted overlay into a container for autoclaving. This also will remove degenerative cells from the plaque areas rendering the plaques more visible.

6. Stain with Wright stain (p. 181) and examine microscopically using a 10X objective. Count the number of plaques and correlate with the amount of virus added initially.

CELLS SUSPENDED IN AGAR

While the monolayer plaque titration methods can be used satisfactorily with many cell lines, in some instances the addition of an overlay medium may cause rounding of the cells and destruction of the monolayer so that plaques are difficult to see. This is particularly true when minute plaques are formed on fibroblast monolayer cultures. To overcome this difficulty a method has been devised to utilize a dense suspension of cells in agar (11).

MATERIALS

1. Twenty ml of liquid 1. 5% Noble agar* in 199 peptone (p. 238). This is prepared by mixing one volume of 3. 0% Noble agar in distilled water with one volume of 2X 199 peptone.

2. Twenty ml of liquid 0. 75% Noble agar in 199 peptone. This is prepared by diluting 10 ml of the above 1. 5% agar with an equal volume of 199 peptone.

3. Water bath adjusted to 48° C.

* Difco Laboratories, Detroit, Michigan.

NOTES

4. One ml of herpes simplex virus diluted in 199 peptone to contain approximately 20-60 plaque forming units (PFU) per ml.

5. Five ml of 199 peptone containing 5×10^6 L-M mouse cells/ml. These cells should be from either monolayer or suspension cultures in the mid-logarithmic growth phase.

6. Six sterile 2 oz. French square bottles with rubber lined screw caps.

7. Six 10 ml and one 1 ml sterile cotton-plugged serological pipettes.

8. Five ml of tetrazolium solution. This consists of 0.75 mg/ml of 2-p-iodophenyl-3 (-p-nitrophenyl) 5-phenyl-tetrazolium chloride in 0.9% NaCl.

PROCEDURE

1. An agar underlay is prepared by transferring 4 ml of 0.75% Noble agar in 199 peptone to each of 5 two oz. French square bottles (use a 10 ml pipette warmed to 50° C). Allow the agar to solidify.

2. With a sterile 1 ml pipette transfer the 1 ml of herpes virus suspension to the tube containing 5 ml of L-M cell suspension. Mix well and decant the virus-cell mixture into a 2 oz. French square bottle containing 6 ml of melted 1.5% Noble agar in 199 peptone (50°C). Swirl to mix.

3. With a warm (50° C) sterile 10 ml pipette take up 5 ml of the virus-cell-agar mixture (step 2) and quickly transfer 1 ml to each of the five two-oz. French square bottles previously prepared (Step 1). Allow the agar containing virus and cells to solidify over the base agar layer.

4. Incubate the bottles at 35°C.

5. After 3-4 days add 1 ml of tetrazolium solution and incubate at room temperature for 2 hours. Count the number of clear plaques and compare with the estimated numbers of virus particles put into the bottles.

NOTES

MICROBIOLOGY

EXERCISE 1. INTRACELLULAR GROWTH OF S. TYPHOSA IN L-M CELLS

Cell culture techniques have proved to be an invaluable aid in the study of intracellular parasitism by certain bacteria. Particular attention has been given to *M. tuberculosis* (14-16), Brucellae (17-19) and Salmonellae (20-22). Monocytes or other actively phagocytic cell types have been employed in many of these studies. However, both epithelial cells (16) and fibroblasts (20) have been demonstrated to phagocytize bacteria *in vitro* and can be more readily maintained in continuous culture than can monocytes.

The method described here is based on a technique used in the Department of Microbiology at the University of Michigan (23). It is similar to procedures described by Showacre, Hopps, *et al* (20). It is designed to demonstrate the principle of intracellular growth of bacteria *in vitro*. Since most bacteria will grow readily in tissue culture media, an important point of technique in the experiment is the use of streptomycin in the medium to prevent extracellular growth of the salmonellae. The antibiotic is added after an initial period of incubation which permits the tissue cells to phagocytize the organisms.

MATERIALS

1. Monolayer culture of L-M mouse cells in medium 199P (p. 238).
2. Twelve sterile screw-cap Leighton tubes containing coverslips.
3. Sterile rubber policeman.
4. Four sterile, cotton plugged, 10 ml serological pipettes.
5. Twenty-five ml sterile $Eagle_{80}$ Horse $serum_{20}$ (p. 237).
6. Twenty-four cotton plugged, sterile, disposable capillary pipettes.
7. Three sterile, cotton plugged, 10 ml serological pipettes.
8. Two sterile, cotton plugged, 1 ml serological pipettes.
9. Twenty-five ml sterile 2X Eagle medium (p. 239) containing 50 µg streptomycin/ml.
10. Twenty-four hour trypticase soy broth* culture of *S. typhi*, strain Ty2. **
11. Twenty five ml of sterile 2X Eagle medium (p. 239) without antibiotics.
12. #1 McFarland barium sulfate turbidity standard (p. 242).
13. Tray of 2% Osyl*** for disinfecting contaminated materials.

NOTE: Since salmonellae are pathogens, appropriate bacteriological techniques must be used. Strain Ty2 <u>is capable of infection</u> and, if used, is due the respect accorded any pathogen.

* Baltimore Biological Laboratories, Baltimore, Md.
** Walter Reed Army Institute of Research.
*** Lehn and Fink Products Corp., Bloomfield, N.J.

NOTES

PROCEDURE

1. Harvest the monolayer of L-M cells and prepare 12 Leighton tube cultures as described in Exercise 1, p. 11. Use $Eagle_{80}$ Horse $serum_{20}$.

2. Incubate the Leighton tubes at 35° C.

3. After 3 days remove the medium from each tube using a capillary pipette and discard the medium.

4. Rinse each tube twice with 2X Eagle medium without antibiotics.

5. Centrifuge the broth culture of *S. typhi*. Discard the supernatant fluid and resuspend the organisms in 2X Eagle medium (without antibiotics) to give a suspension containing approximately 3×10^8 organisms/ml as determined by comparison with a #1 barium sulfate turbidity standard.

6. Add 0. 1 ml of the bacterial suspension to each of the washed monolayers and incubate at 35° C for 3 hours.

7. After 3 hours remove the inoculum and discard into 2% Osyl. Wash the monolayers twice with 1 ml of 2X Eagle medium containing 50 μg of streptomycin. Discard the washes in Osyl.

8. Add 1 ml of 2X Eagle medium containing 50 μg of streptomycin to each of 10 of the cultures. Remove the coverglasses from the two remaining tubes, fix and stain with May-Grünwald Giemsa (p. 182). These coverslips will serve as the control preparations at zero hours incubation.

9. Incubate the remaining 10 tubes at 35°C. Remove coverglasses from two tubes each day for the next five days. Fix and stain as directed under 8 above. If the pH of the medium in any of the cultures falls below 7. 2 readjust by adding $NaHCO_3$.

10. Examine the stained preparations with the oil immersion objective and count the number of organisms per cell in 100 cells. Tabulate the data needed to provide evidence of intracellular multiplication.

NOTES

IMMUNOLOGY

The study of tissue cell antigens has become increasingly popular in such areas as tissue transplantation, tumor immunology and immunologic disease. In addition, it appears that serologic specificity is the most reliable criterion of species identity of cells in culture (24-26).

Both cell and organ culture techniques also have been applied to studies of antibody formation. The secondary antibody response has been investigated extensively (27) and encouraging results have been obtained in efforts to initiate antibody synthesis *in vitro* (28).

Several methods for titrating tissue cell antisera have been employed. Those which have been applied most commonly to tissue culture materials are hemagglutination (24), immunofluorescence (25), mixed agglutination (26), cytotoxicity (29-30) and direct agglutination.

NOTES

EXERCISE 1. TITRATION OF SPECIFIC TISSUE CELL ANTISERUM BY CYTOTOXICITY AND DIRECT AGGLUTINATION.

The method described here measures both cytotoxicity and agglutination. By utilizing vinyl cup panels and incorporating phenol red it is possible to get a sensitive measure of antibody with small amounts of material.

MATERIALS

1. Two tenths ml of sterile, heat inactivated (60°C for 30 min.) rabbit antiserum* against L-M mouse fibroblasts.

2. Two tenths ml of sterile, heat inactivated normal rabbit serum (pre-immune).

3. Two and one half ml of sterile, unheated, fresh, normal guinea pig serum (source of complement).

4. Two and one half ml of sterile, heated (60°C for 30 min.) normal guinea pig serum.

5. Monolayer culture of L-M mouse cells in 199P.

6. One hundred ml of sterile 199P (p. 238).

7. Sterile vinyl cup panels** (4x11 depressions). These panels may be sterilized by soaking for at least 2 hours in 95% ethanol. Each panel is then grasped at one corner with sterile forceps and the excess alcohol is allowed to drain off after which the panel is dropped into a sterile paper bag, sealed and allowed to sit for 12-18 hours.

8. Template to hold vinyl cup panels. This can be made by drilling holes in a piece of 1/8 inch "Masonite" board so that the panel rests in the holes. One half inch legs should be fastened to each corner.

9. Two inch wide roll of "Scotch" cellophane tape.***

10. Twenty sterile 16 X 125 mm screw cap culture tubes.

11. Sterile 4 oz. prescription bottle.

* Isoantibodies, which are known to be present in rabbit sera, have not been found to interfere with this procedure.

** Fabri-Kal Corporation, Kalamazoo, Michigan.

*** Minnesota Mining and Manufacturing Co., Minneapolis, Minnesota.

NOTES

12. Sterile rubber policeman.

13. Forty-two sterile, cotton plugged 1 ml serological pipettes.

14. Four sterile, cotton plugged 10 ml serological pipettes.

PROCEDURE

1. Prepare a consecutive series of 10 twofold dilutions of rabbit antiserum against L-M strain cells in 199 peptone as follows: Pipette 1.8 ml of 199 peptone into the tube containing 0.2 ml of antiserum and 1.0 ml of 199 peptone into each of 9 additional tubes. With a 1 ml pipette mix the contents of the tube containing antiserum and transfer 1 ml of the mixture to the second tube. Discard the pipette and with a fresh pipette mix the contents of tube 2 and then transfer 1 ml of the mixture to tube 3. Repeat these twofold dilutions through tube number 10. Discard 1 ml of the mixture in tube 10.

2. Prepare a similar series of dilutions of normal (pre-immune) rabbit serum. Adjust the pH of all dilutions to 7.4.

3. Harvest the L-M mouse cells by scraping into fresh medium 199P (p. 238). Triturate to homogenize the suspension and dilute to the concentration specified for you by the instructor ($2X10^4$/ml, $4X10^4$/ml, $6X10^4$/ml or $8X10^4$/ml). Prepare at least 50 ml of cell suspension. Adjust to pH 7.4 if necessary.

4. Transfer 0.5 ml of the antiserum dilutions to cups 1-10 of the first two rows on the vinyl cup panel so that 0.5 ml of the contents of tube 1 goes into cup 1 of row 1 and 0.5 ml into cup 1 of row 2, etc. Transfer the dilutions of normal serum to cups 1-10 of rows 3 and 4 in the same manner. Add 0.5 ml of 199 peptone to cup 11 in each row as a control.

5. Add 0.5 ml of cell suspension to each of the 44 cups. Take care to ensure that the suspension is adequately mixed so that replicate samples are delivered to the cups. A Cornwall syringe* is of value for this step.

6. Add 0.1 ml of unheated guinea pig serum to each cup in rows 1 and 3. Add 0.1 ml of heated guinea pig serum to each cup in rows 2 and 4.

7. Seal each two rows of the vinyl cup panel with 2 inch "Scotch" tape and incubate the panel at 35°C.

* Becton, Dickinson Co., Rutherford, New Jersey.

NOTES

8. Examine the cultures each day for shift of pH (as indicated by change of color of phenol red in the medium) and for growth (microscopic examination). The panel may be inverted on the microscope stage and the cells examined on the bottom of the cups. Also record evidence of agglutination and/or lysis. Read and record end points of (a) agglutination, (b) lysis, (c) metabolic inhibition. Cells which are not inhibited should gradually lower the pH. Where cells are inhibited or lysed, a rise in pH will result from release of basic compounds.

NOTE: In steps 1-7 it is imperative that you work rapidly to avoid shifts in pH due to loss of CO_2.

Selected students will carry out the above procedure using HeLa cells in place of L-M cells to demonstrate specificity. A monolayer of HeLa cells will be harvested with versene solution (p. 241), centrifuged and resuspended in Eagle $medium_{80}$, calf $serum_{20}$. All dilutions will be made using $Eagle_{80}$ human $serum_{20}$. Otherwise the procedure will be exactly the same.

NOTES

EXERCISE 2. DETERMINATION OF SPECIES OF ORIGIN OF CELLS IN TISSUE CULTURE BY MIXED AGGLUTINATION.

With the demonstration of cross contamination (31) of established cell lines, confirmation of the species of origin has become an important consideration. Several serological procedures have been used to test for species antigens (24-26). The mixed agglutination reaction is based on the fact that species specific antigens are shared widely by cells in the animal body including the erythrocytes. However, mixed agglutination has a wider application than just identification of species (32-34).

MATERIALS

1. Monolayer culture of L-M strain cells in 199 peptone (p. 11).
2. Monolayer culture of HeLa cells in $Eagle_{80}$ human $serum_{20}$ (p. 11).
3. Sterile rubber policeman.
4. Ten ml of sterile versene solution (p. 241).
5. Two 12 ml sterile screw-cap conical centrifuge tubes.
6. Five sterile, cotton plugged 10 ml serological pipettes.
7. One hundred ml of diluent (0.02% crystalline bovine plasma albumin in 0.9% NaCl) in a 250 ml Erlenmeyer flask.
8. Fifty siliconed Pasteur pipettes.
9. Thirty siliconed 6 x 60 mm tubes.
10. Fifteen siliconed 2 x 50 mm tubes.
11. Fifteen siliconed melting point tubes.
12. Ice bath to hold a 48-tube rack and flask of diluent.
13. Mixture of paraffin and petrolatum (3:1) and camel hair brush.
14. Fifteen siliconed 22 mm square #2 coverglasses.
15. Six siliconed microscope slides.
16. Five 2 ml rubber dropper bulbs (one with rubber insert to fit melting point tubes).
17. One half ml of freshly drawn mouse blood in Alsever solution (p. 243).
18. One half ml of freshly drawn human group O blood in Alsever solution.
19. One tenth ml of rabbit anti-mouse erythrocyte antiserum absorbed with human group A erythrocytes to remove anti-human antibody.
20. One tenth ml of rabbit anti-human group O erythrocyte antiserum absorbed with mouse erythrocytes to remove anti-mouse antibody.

PROCEDURE

1. Set up an ice bath with a rack to hold 6 X 60 mm tubes and to contain an Erlenmeyer flask with 100 ml of diluent.

2. Prepare 100 ml of diluent and place the flask in the ice bath.

NOTES

3. Inactivate the antisera at 60°C for 30 minutes and store in ice bath.

4. Harvest the L-M cells into the used medium, by scraping, and triturate to disperse (p. 12).

5. Remove the medium from the HeLa culture, add 10 ml of versene and incubate at 35°C for 15-20 minutes. Dislodge the cells and triturate.

6. Centrifuge the L-M and Hela cells at 150G, wash twice with diluent and resuspend in diluent to a concentration of 20 cells/40X microscopic field (40X objective). Use siliconed Pasteur pipettes and siliconed tubes! Store in the ice bath until needed.

7. Wash the mouse and human erythrocytes three times with the cold diluent and resuspend in diluent to a concentration of 0.5% (by volume). Centrifuge at 500G to pack the cells at each wash. Store in the ice bath.

8. Place 12 siliconed 6 X 60 mm tubes in a rack and number them from 1-12. Place 2 drops of L-M cell suspension (washed and adjusted to proper concentration) in each of the first six tubes. Place 2 drops of Hela cell suspension in each of the last six tubes.

9. Prepare a 1:10 dilution of each of the antisera using the cold diluent. Add 2 drops of anti-mouse erythrocyte antiserum to tube 1, tube 2, tube 7 and tube 8. Add 2 drops of anti-human erythrocyte antiserum to tube 3, tube 4, tube 9 and tube 10. Add 2 drops of saline to tubes 5, 6, 11, 12. Mix well!

10. Allow the tubes to incubate at room temperature (28°C) for 1 hour with frequent agitation (or on a slowly rotating drum).

11. After 1 hour centrifuge the tubes at 150 G and wash the sensitized cells three times with diluent. Use care in pipetting off supernatant to avoid discarding cells.

12. Resuspend the cells in one drop of diluent. Label twelve 2 X 50 mm tubes 1-12 and transfer the drop of cell suspension to the corresponding tube. Care must be exercised to avoid losing the cells at this point. Use siliconed Pasteur pipettes.

13. Add one drop of mouse erythrocyte suspension to tubes 1, 3, 5, 7, 9, and 11. Add 1 drop of human erythrocyte suspension to tubes 2, 4, 6, 8, 10 and 12.

NOTES

14. Centrifuge for 2 minutes at 150G.

15. Divide each of six 1X3 inch microscope slides in half using a marking pen. Number the sections 1-12.

16. Using a separate siliconed melting point tube for each tube of the test, carefully take up a drop of the deposited cell mixture and transfer it to the corresponding slide section. Use the rubber bulb with the rubber insert.

17. Cover each drop with a siliconed coverslip and seal with the paraffin-petrolatum mixture using a camel hair brush.

18. Examine the slides microscopically using a 10 X ocular and a 40 X objective. If possible utilize phase contrast optics. Evaluate the reaction in terms of the numbers of tissue cells (L-M or HeLa) demonstrating mixed agglutination as well as in terms of the average numbers of indicator cells per reacting tissue cell.

    NOTE: Positive reactions will vary from tissue cells which are completely covered with indicator cells to those with only a few indicator cells stuck to the surface. In the case of weak reactions it may be necessary to tap the coverslip gently while watching the slide to verify that the indicator cells are actually attached to the tissue cells.

19. Prepare a table as indicated on page 120 and record your results as ++++ (majority of tissue cells showing strong reactivity) to + (only few cells reactive and few indicator cells per reacting-tissue cell). Numbers 5, 6, 11 and 12 are controls for nonspecific sticking of indicator cells to the tissue cells. If any of these controls are positive the corresponding tests must be considered to be invalid.

NOTES

Tissue Cells	Antiserum	Mouse erythrocytes			Human erythrocytes		
L-M Mouse Cells	Rabbit antimouse RBC 1:10	M	I	T ①	M	I	T ②
	Rabbit antihuman RBC 1:10	M	I	T ③	M	I	T ④
	Diluent Control	M	I	T ⑤	M	I	T ⑥
Hela Cells	Rabbit antimouse RBC 1:10	M	I	T ⑦	M	I	T ⑧
	Rabbit antihuman RBC 1:10	M	I	T ⑨	M	I	T ⑩
	Diluent Control	M	I	T ⑪	M	I	T ⑫

M = mixed agglutination

I = Agglutination of indicator cells only!

T = Agglutination of tissue cells only!

NOTES

ENDOCRINOLOGY

EXERCISE 1. EFFECTS OF TRI-IODOTHYRONINE ON CELLS

Although the gross effects of hormones and their analogs on organ systems *in vivo* has been reasonably well characterized, little is known of hormone function at the cellular level because of the complexity of tissue organization and interaction. Isolation of cells in culture systems affords a means for study of hormone action as it may relate to: (a) permeability and transport; (b) mechanisms and rate of cell growth; (c) morphologic modulation and differentiation; (d) preferential use of alternative biochemical pathways; (e) genetic constitution of cells and its variation; (f) susceptibility of cells to viral infection; (g) antibody synthesis; *etc.*

The effect of tri-iodothyronine on cells will be surveyed in this exercise by application of the clonal plating technique. The experimental approach used here is applicable generally to study of other hormones or their analogs.

MATERIALS

1. Monolayer of HeLa cells, log phase of growth.

2. Fifteen ml of growth medium, Eagle medium 80%, human serum 20% (p. 237).

3. Five sterile 2 oz bottles, French square with rubber lined screw caps.

4. Sterile trypsin solution, 0. 2% (p. 240).

5. Five sterile 16 x 125 mm screw-cap tubes each containing 4. 5 ml of sterile $Eagle_{80}$ human $serum_{20}$ (p. 237). These are dilution blanks.

6. Growth medium containing tri-iodothyronine* divided in aliquots to give a final concentration when diluted 1:3 of 50, 75, 100 and 125 µg/ml respectively. The procedure for preparation of solutions of tri-iodothyronine is described by Murphy, *et al.* (35).

7. Eight sterile cotton plugged 5 ml serologic pipettes.

* California Corporation for Biochemical Research, Los Angeles, California.

NOTES

PROCEDURE

1 - 3. See steps 1-3 page 54.

4. Dispense 150 cells contained in a volume of 2 ml of growth medium into each of 5 screw-cap culture vessels.

5. Add to each of four of these culture vessels 1 ml of growth medium containing tri-iodothyronine to give final concentrations per ml respectively of 50, 75, 100 and 125 μg. Vessel five receives 1 ml of growth medium free of tri-iodothyronine and serves as control. Incubate the cultures at 35°C.

6. Observe the bottles daily for a period of ten days and record the following data in a suitable table:

   a. Relative plating efficiency (p. 52) of cells in each culture vessel compared with the control.

   b. Effect of the drug on the morphology of cells.

   c. Effect of the drug on production of acid by cells.

   d. Time required for cells to form a monolayer in each vessel.

7. Plot a dose-response curve in which toxicity (dependent variable) scored in terms of 0, 1+, 2+, 3+ and 4+ (p. 90) degeneration is plotted against dose of tri-iodothyronine (abscissa). Plots of cell growth rate, plating efficiency, acid production, *etc.*, can be made similarly.

NOTES

EXERCISE 2. EFFECT OF PREDNISOLONE ON ALKALINE PHOSPHATASE CONTENT OF CELLS.

As demonstrated in the previous exercise (p. 124) the culture environment to which cells are exposed can dramatically alter their metabolism. For example, the addition of Prednisolone (a hydrocortisone analogue) can be shown to markedly increase their alkaline phosphatase activity (36, 37).

MATERIALS

1. Monolayer culture of HeLa cells adapted to 199 and 10% human serum.
2. Ten ml sterile 0.05% trypsin (p. 240).
3. Four sterile screw-cap Leighton tubes each containing one 9x22 mm #1 coverglass.
4. Ten ml calcium magnesium-free phosphate buffered saline (CMF-PBS).
5. Five ml sterile 199 with 10% human serum (control medium).
6. Five ml sterile 199 with 10% human serum containing 1.0 μg. Prednisolone* per ml. (experimental medium).
7. Cold neutral buffered formalin fixative.
8. Reagents for the histochemical demonstration of alkaline phosphatase (p. 194).
9. Five sterile cotton plugged 10 ml and four 1 ml serological pipettes.
10. One sterile 15 ml conical centrifuge tube.
11. Thirty ml of Hanks BSS (p. 214).

PROCEDURE

1. Pipette medium off of stock cells.
2. Wash once with an equal volume of CMF-PBS.

* Δ^1- hydrocortisone phosphate. Merck, Sharpe and Dohme.

NOTES

3. Replace with 10 ml of 0.05% trypsin solution. Incubate at room temperature until cells are detached.

4. Triturate and transfer to 15 ml centrifuge tube. Centrifuge at 500 RPM for 5 minutes. Decant off trypsin solution and resuspend cells in 10 ml of 199 containing 10% human serum.

5. Count cells and seed each of the four tubes with 1×10^6 cells in 0.5 ml. of medium.

6. Add the control medium (199 plus human serum) to two of the tubes to bring final volume of each to 1 ml. Add experimental medium (199, human serum and prednisolone) to the remaining two tubes.

7. Incubate at 35°C.

8. When cells appear confluent (several days), remove coverslip, wash with three changes of BSS and fix in cold neutral buffered formalin.

9. Stain coverslip for alkaline phosphatase (p. 194).

10. Examine slides and compare the control cells and experimental cells as to alkaline phosphatase activity, cell number, cell size, and general morphology.

NOTES

EXERCISE 3. EFFECT OF TESTOSTERONE PROPRIONATE ON PROSTATE CULTURES.

The organ culture technique provides an effective means of determining the direct effects of a hormone on the morphology and metabolism of target organs. In recent years this culture method has been used to demonstrate the action of insulin (38, 39), thyroxine (40-43), steroids (44-48), and trophic hormones of the anterior hypophysis (49-53) among others. The particular exercise outlined below exemplifies the direct action of a male hormone, testosterone proprionate, on its sensitive target organ, the prostate (54). In this procedure, the plasma clot technique is proposed although the raft technique (p. 76) would be equally useful.

MATERIALS

1. One male mouse, 4-6 weeks of age.
2. Twelve sterile plastic organ culture dishes. *
3. Seven ml chicken plasma (p. 203).
4. Seven ml chicken plasma to which 0. 5 mg testosteron proprionate has been added.
5. Five ml embryo extract$_{50}$ (p. 205).
6. Six sterile cotton plugged 5 ml serological pipettes.
7. Fifteen sterile cotton plugged capillary pipettes.
8. Two sterile scalpels (Iris or Bard-Parker #10).
9. Two sterile fine forceps (Iris or Jewelers).
10. Thirty ml sterile Hanks BSS.
11. Thirty ml sterile distilled water.

PROCEDURE

1. Add 9 drops of chicken plasma to the central concavity of each of two organ culture dishes followed by 4 drops of EE_{50}. Quickly but gently mix before medium clots (control medium).

2. Repeat step one, using the plasma containing testosterone proprionate (final concentration will be approximately 50 μg/ml).

3. Add 5 ml of sterile distilled water to the absorbent paper in the outer ring of each of the four dishes.

* Falcon Plastics, Los Angeles, California.

NOTES

4. Aseptically remove the ventral prostates from the mouse.

   This is accomplished as follows:

   a. Kill the mouse by cervical dislocation and pin him out on a cork board, dorsal surface down.

   b. Swab the lower abdominal region with 70% ethanol and make a single longitudinal incision from abdomen to the external genitalia. Carefully separate the skin and underlying abdominal muscle and identify the bladder lying midventrally in the pelvic region. Using sterile forceps lift the bladder and note the paired ventral lobes lying on and to the side of the urethra, just caudal to the base of the bladder. Carefully separate each lobe from the surrounding fascia and cut away from the urethra. Place in a petri dish containing BSS.

5. Cut explants from each lobe to approximately 2 mm^3 and transfer to surface of plasma clot with a capillary pipette. Use explants from one ventral prostate for control dishes and the other side for the media with hormone. Remove any excess BSS from around the explants with the capillary pipette.

6. Incubate at 35° C for about seven days, subculturing every second or third day (p. 74).

7. Fix in Bouin (p. 178), section and stain in hematoxylin and eosin (p. 185).

8. Examine sections for prostatic epithelium. Note differences in cell morphology, height of epithelial layer and size of alveoli.

NOTES

NUTRITIONAL STUDIES

EXERCISE 1. APPLICATION OF THE CLONAL TECHNIQUE TO NUTRITIONAL STUDIES OF PRIMARY AND ESTABLISHED CELL CULTURES.

The clonal technique (55) for study of cells is one of a number of available procedures which can be employed to provide information concerning growth characteristics and nutritional requirements of cells (56-58). The clonal procedure for studying the nutritional requirements of cells depends upon evaluation of the relative plating efficiency of cells in media of different composition. The technique may fail to discriminate between those constituents of media required for attachment of cells to glass surfaces and those required for sustained optimal growth of cell strains on serial subculture.

MATERIALS

1. One bottle of log phase HeLa cells grown in $Eagle_{80}$ human $serum_{20}$ (p. 11).
2. Five ml of sterile trypsin solution, 0.2% (p. 240, stock solution).
3. Twenty ml of Eagle $medium_{80}$ human $serum_{20}$ (p. 237).
4. Two ml of sterile human serum and two ml of horse serum.
5. Haemocytometer.
6. Ten sterile 4 oz, French-square bottles with rubber lined screw caps.
7. Ten sterile 16 x 125 mm screw-cap tubes for preparation of serum-Eagle medium mixtures.
8. Five sterile 16 x 125 mm screw-cap tubes each containing 4.5 ml $Eagle_{80}$ human $serum_{20}$.
9. Five each of sterile cotton plugged 5 ml and 10 ml serologic pipettes.
10. Fifteen sterile cotton plugged 1 ml serological pipettes.

PROCEDURE

1. Prepare a mondisperse suspension of HeLa cells in complete medium (p. 42 - steps 1-3) to give 10 ml of cell suspension (final concentration of approximately 450 cells per ml).

2. Prepare 2 ml of each of the following media in a culture vessel:

Human serum mixtures	Horse serum mixtures
Eagle medium 80% human serum 20%	Eagle medium 80% horse serum 20%
" " 85% " " 15%	" " 85% " " 15%
" " 90% " " 10%	" " 90% " " 10%
" " 95% " " 5%	" " 95% " " 5%
" " 99% " " 1%	" " 99% " " 1%

NOTES

3. Pipette 1 ml of cell suspension into each culture vessel.

4. Incubate cells at 35°C for 10 days; observe cells daily for attachment and growth.

5. Stain the clones of cells *in situ* with Wright stain (p. 181) and record the distribution of different morphologic types of colonies in the various culture vessels (p. 56).

6. Evaluate the relative nutritional adequacy of the test sera by plotting relative plating efficiency against per cent serum in the medium. (The bottle containing Eagle medium 80%, human serum 20% is the control.)

NOTES

EMBRYOLOGY

EXERCISE 1. HISTOGENESIS OF BONE

Cell and organ culture have contributed significantly to our present knowledge of embryogenesis and histogenesis. Space does not permit consideration of the many areas under investigation (See reviews 59-64). The following experiment (65-66) is included as representative of how organ culture can be used to study histogenic processes.

MATERIALS

1. Two sterile organ culture dishes (plastic) and two stainless steel rafts*.
2. Two sterile forceps (Iris).
3. Two sterile scalpels (Iris or Bard-Parker #10).
4. Five ml sterile medium [Biggers BGJ_b (67) Waymouth' s (p. 229) or 1066**].
5. Five sterile disposable capillary pipettes.
6. Ten ml sterile distilled water.
7. Two five ml cotton plugged serological pipettes.
8. One chick embryo 7-10 days incubation.
9. Three sterile Petri dishes.
10. Twenty-five ml sterile Hank BSS (p. 214).

PROCEDURE

1. Moisten outer disc of organ culture dish with 5 ml. distilled water.

2. Aseptically place a stainless steel raft over center well.

3. Aseptically remove embryo (p. 205) and transfer to sterile Petri dish.

4. Cut off lower limbs as close to body wall as possible and carefully remove all tissue adhering to the cartilaginous tibia-tarsus.

5. Transfer bones to another sterile Petri dish and wash in sterile BSS.

6. Transfer the bones to the stainless steel raft.

* Falcon Plastics, Los Angeles, California.

** Connaught Medical Research Laboratory, Toronto, Canada.

NOTES

7. With a sterile capillary pipette, add sufficient medium to the center well until it just touches the bottom of the raft.

8. Incubate at 35° C.

9. Subculture at 3 day intervals. Carefully remove out-growth and transfer bones to a new raft with fresh medium.

10. Check the cultures microscopically every day and make camera lucida drawings as a permanent record.

11. Fix and stain for histologic examination after 6-7 days *in vitro* (pp. 178, 185).

NOTES

EXERCISE 2. TISSUE INTERACTIONS *IN VITRO*

The growth stimulating and organizing action shown by certain tissues on neighboring cells and tissues in the development of the early embryo (induction) has been a recognized phenomenon for some time. Grobstein (68, 69) has recently provided experimental evidence to demonstrate that embryonic induction reflects the transmission of a substance (s) between the two interacting tissues.

By the use of a membrane filter interposed between two such tissues (embryonic spinal cord and embryonic metanephros), induction may be demonstrated if the thickness of the membrane filter is not too great.

Grobstein's basic technique, which is presented below in a modified form, can be a generally useful procedure for studying tissue interactions.

MATERIALS

1. One mouse, pregnant 11 days.
2. Ten sterile Petri dishes.
3. Two sterile fine forceps.
4. Two sterile iridectomy knives.
5. One organ culture dish (plastic). *
6. One fine scissors.
7. One supported ring assembly**.
8. Twenty ml sterile horse $serum_{50}$ BSS_{50}.
9. Twenty ml sterile Hanks balanced salt solution.
10. One ml sterile chicken plasma.
11. Three ml sterile Grobstein's media (Horse $serum_{40}BSS_{40}$Embryo Extract (9 $days)_{20}$).
12. Ten ml sterile 3% Trypsin solution in CMF-PBS.
13. 70% ethanol.
14. Three sterile capillary pipettes.

PROCEDURE

1. Sterilize supported ring assembly by soaking in 70% ethanol for thirty minutes. With a sterile forceps, remove assembly, allow excess alcohol to drain off and store assembly in a sterile petri dish overnight.

* Falcon Plastics, Los Angeles, Calif.

** A rectangular piece of acrylic plastic, 2 cm x 6 mm x 3/32" with a 3 mm hole bored in center. A piece of membrane filter (Millipore Co.: Type THWP, pore size 0.47u, 25u thick) is glued over the hole with cement made from a piece of membrane filter dissolved in acetone to give a syrupy fluid.

NOTES

2. Just before use, soak the supported ring assembly in a 1:1 mixture of horse serum and BSS and allow to remain in this solution until ready to use.

3. Aseptically remove the uteri from a mouse which is about 11 days pregnant. Aseptically transfer uteri to a sterile petri dish containing balanced salt solution.

4. Dissect open the uteri and carefully transfer the embryos, one at a time, to separate sterile Petri dishes containing BSS and horse serum (1: 1).

5. Separate the caudal part of the body (include hind limbs, cloacal region and tail) of each embryo. Save the cephalic end for spinal cord source (step 12).

6. Using a fine iridectomy knife, cut just ventral to the spinal cord in order to separate the dorsal part from the rest of the body wall.

7. Lay the resulting ventral body wall with the dorsal cut surface down and examine with a strong source of transmitted light.

8. Carefully excise the paired metanephri located on either side of the cloaca with fine iridectomy knives.

9. Transfer the trimmed rudiments to the trypsin solution for 3-5 minutes.

10. By carefully triturating the tissue with a capillary pipette, the epithelial component should separate from the mesenchymal components.

11. Transfer the metanephric mesenchyme to a small amount of horse serum - BSS (1:1).

12. Using the remaining cephalic part of the embryo, carefully cut ventral to the spinal cord with fine iridectomy knives. Remove the cervical and thoracic part of the spinal cord.

13. With the aid of a dissecting microscope, cut away the neighboring tissue and ganglia from the spinal cord with a pair of fine iridectomy knives.

14. Split the spinal cord in a frontal plane so as to separate the thinner dorsal part of the cord from the thicker ventral part. Cut the dorsal part in 0.2 to 0.4 mm explants.

NOTES

15. Place the sterilized and pre-soaked supported ring assembly into the plastic organ culture dish with the cup side up.

16. Add one piece of spinal cord to the well of the cup. Remove excess fluid and add two drops of an equal mixture of plasma and Grobstein's medium (freshly mixed). Allow to clot.

17. Invert the assembly and place two pieces of metanephric mesenchyme directly on the membrane filter, one over the area of the spinal cord. The second should be placed at least 1 mm away. Remove excess fluid and add equal mixture of plasma and medium. Allow to clot.

18. Fill the well of the organ culture dish with sufficient medium so that the fluid just reaches the top of the ring assembly.

19. Incubate in 5% carbon dioxide in air for three days. The atmosphere should be saturated with water by adding water to the outer ring of the organ culture dish.

20. After three days incubation, fix the membrane and associated explants in Bouin (p. 178). Section at right angles to the filter.

NOTES

PHARMACOLOGY

EXERCISE 1. CELL CULTURE FOR CHEMICAL SCREENING

The cell culture technique has received a great deal of emphasis in recent years in cancer chemotherapy screening. Since this assay procedure is rapid, reproducible and relatively inexpensive, the Cancer Chemotherapy National Service Center (CCNSC) has designated cell culture as a prescreen for the standard mouse tumor system (70). Although the correlation between the cytotoxicity of chemical agents on mouse tumors when compared with cell cultures is not perfect (71-72), the systematic use of both procedures has proved extremely valuable.

The procedure described below is modified from Toplin (73) and utilizes graded doses of test solutions in disposable plastic cup panels, followed by the addition of a standard aliquot of a HeLa cell suspension and microscopic evaluation after a defined period of culture.

MATERIALS

1. One half mg sterile crystalline Actinomycin D. *
2. One gm sterile crystalline Choramphenicol. **
3. Stock monolayer culture of HeLa cells in log-phase of growth (p. 11).
4. Fifty ml of $Eagle_{80}$ Human $serum_{20}$ (p. 237).
5. Two sterile vinyl cup panels*** (2 x 6 depressions). Sterilize by immersing in 95% ethanol for two hours, drain and store in a sterile paper bag at least overnight.
6. Two inch "Scotch" cellophane tape****.
7. Fifteen sterile 16 x 125 mm screw-cap culture tubes.
8. Sterile 3 oz prescription bottle.
9. Fifteen sterile 1 ml cotton plugged serological pipettes.
10. Two sterile disposable 5 ml syringes with #25 needles.
11. Ten sterile screw-capped 16 x 125 mm tubes each containing 4. 5 ml $Eagles_{80}$ Human $serum_{20}$.
12. One template for vinyl cup (see 8. , p. 108).

* Drug Control Section, Cancer Chemotherapy, NSC, 7981 Eastern Avenue, Silver Springs, Md.

** Parke, Davis and Co., Detroit, Michigan.

*** Fabri-Kal, Kalamazoo, Michigan.

**** Minnesota Mining & Manufacturing, Minneapolis, Minnesota.

NOTES

PROCEDURE

1. Add 1.0 ml sterile cold distilled water to the sterile crystalline Actinomycin D. Adjust pH to 7.2 - 7.4 if necessary.

2. Prepare consecutive ten-fold serial dilutions by adding 0.5 ml of the Actinomycin D solution to one of the tubes containing 4.5 ml of $Eagle_{80}$ Human $serum_{20}$. (50 μg/ml). Repeat for a total of 5 dilutions (5 μg, 0.5 μg, 0.05 μg, 0.005 μg).

3. Add 5.0 ml sterile distilled water to the sterile crystalline Chloramphenicol. Adjust to pH 7.2-7.4 if necessary. Prepare 5 similar series of ten-fold dilutions (20 mg, 2 mg, 0.2 mg, 0.02 mg, 0.002 mg/ml).

4. Harvest HeLa cells (p. 12) and resuspend in $Eagle_{80}$ Human $serum_{20}$. Count (p. 156) and dilute to 30,000 to 40,000 cells/ml Prepare at least 15 ml of cell suspension. Adjust to pH 7.4 if necessary.

5. Place panel in template.

6. Place 0.5 ml of each ten-fold dilution of Actinomycin D in each of two cups, add 0.5 ml of Hanks BSS into remaining two cups of the panel.

7. Repeat for chloramphenicol using a second panel tray.

8. Add 0.5 ml of cell suspension to each of the twenty-four cups.

9. Seal with the two inch "Scotch" tape and incubate at 35° C.

10. After 5 days of incubation, examine cups for a change in pH and microscopically for growth using 40 X objective. Using the following scale, rate each cup for degree of lysis.

    0 = cells firmly attached and in complete monolayer.
    1 = growth inhibited, some rounding of cells, most adhering.
    2 = cells suspended and clumped. Loss of cellular detail.
    3 = single suspended cells and considerable cell debris.
    4 = complete cytolysis, all debris.

11. Record endpoints for each dilution and determine the cytotoxic endpoint (CE) defined as greater than an average rating of 1 over control ratings, and the lethal endpoint (LE) defined as an average of 3 or better.

NOTES

REFERENCES

1. WHITELOCK, O.V. St., (ed.), The cytopathology of virus infection, Ann. New York Acad. Sc., 81:1-213 (1959).
2. FELTON, H. M. (ed.), Host-parasite Relationships in Living Cells, Springfield, Ill., Charles C. Thomas, (1957).
3. American Public Health Association, Diagnostic Procedures for Virus and Rickettsial Diseases, 2nd ed., New York, American Public Health Association (1956).
4. REED, L. J., and MUENCH, H., A simple method of estimating fifty per cent endpoints, Am. J. Hyg., 27:493-497 (1938).
5. KÄRBER, G., Beitrag zur kollektiven Behandlung pharmakologischer Reihenversuche, Arche. exper. Path. u. Pharmakol., 162: 480-483 (1931).
6. CHANG, S. L., BERG, G., et al., Application of the "most probable number" method for estimating concentrations of animal viruses by the tissue culture technique, Virology, 6: 27-42 (1958).
7. SALK, J. E., LEWIS, L. J., et al., The use of adjuvants to facilitate studies on the immunologic classification of poliomyelitis viruses, Am. J. Hyg., 54: 157-173 (1951).
8. RIGHTSEL, W. A., SCHULTZ, P., et al., Use of vinyl plastic containers in tissue cultures for virus assays, J. Immunol., 76: 464-474 (1956).
9. GINSBERG, H. S., GOLD, E., et al., Tryptose phosphate broth as supplementary factor for maintenance of HeLa cell tissue cultures, Proc. Soc. Exper. Biol. & Med., 89: 66-71 (1955).
10. DULBECCO, R., and VOGT, M., Plaque formation and isolation of pure lines with poliomyelitis viruses, J. Exper. Med., 99: 167-182 (1954).
11. COOPER, P. D., An improved agar cell-suspension plaque assay for poliovirus. Some factors affecting efficiency of plating, Virology, 13: 153-157 (1961).
12. TYTELL, A. A., and NEUMAN, R. E., A medium free of agar, serum and peptone for plaque assay of herpes simplex virus, Proc. Soc. Exper. Biol. & Med., 113: 343-346 (1963).
13. GIFFORD, G. E., and SYVERTON, J. T., Replication of poliovirus in primate cell cultures maintained under anaerobic conditions, Virology, 4: 216-223 (1957).
14. SMYTH, H. F., The reactions between bacteria and animal tissues under conditions of artificial cultivation. IV. The cultivation of tubercle bacilli with animal tissue in vitro. J. Exper. Med., 23: 283-291 (1916).
15. SUTER, E., Multiplication of tubercle bacilli within mononuclear phagocytes in tissue cultures derived from normal animals and animals vaccinated with B.C.G., J. Exper. Med., 97: 235-245 (1953).
16. SHEPARD, C. C., A comparison of the growth of selected mycobacteria in HeLa, monkey kidney, and human amnion cells in tissue culture, J. Exper. Med., 107: 237-246 (1958).
17. POMALES-LEBRÓN, A., and STINEBRING, W. R., Intracellular multiplication of Brucella abortus in normal and immune mononuclear phagocytes, Proc. Soc. Exper. Biol. & Med., 94: 78-83 (1957).
18. HOLLAND, J. J., and PICKETT, M. J., A cellular basis of immunity in experimental Brucella infection, J. Exper. Med., 108: 343-360 (1958).
19. RICHARDSON, M., and HOLT, J. N., Synergistic action of streptomycin with other antibiotics on intracellular Brucella abortus in vitro. J. Bact. 84: 638-646 (1962).
20. SHOWACRE, J. L., HOPPS, H. E., et al., Effect of antibiotics on intracellular Salmonella typhosa. I. Demonstration by phase microscopy of prompt inhibition of intracellular multiplication. J. Immunol., 87: 153-161 (1961).
21. HOPPS, H. E., SMADEL, J. E., et al., Effect of antibiotics on intracellular Salmonella typhosa. II. Elimination of infection by prolonged treatment, J. Immunol., 87: 162-174 (1961).
22. KOURANY, M., A Study of Interaction Between a Human Monocytic Cell Strain and Salmonella Typhosa. Ph.D Thesis, Univ. of Mich. (1962).
23. SOMERS, K., A Preliminary Study of the Interaction of Salmonella Typhosa in HeLa Cell cultures With a Preliminary Comparison Between S. Typhosa Grown on Bacteriological Medium and in Cell Cultures. M.S. Thesis, Univ. of Mich. (1962).
24. BRAND, K. G., and SYVERTON, J. T., Results of Species-Specific Hemagglutination Tests on "Transformed", Nontransformed, and Primary Cell Cultures, J. Nat. Cancer Inst., 28: 147-157 (1962).

25. STULBERG, C. S., SIMPSON, W. F., et al., Species-related antigens of mammalian cell strains as determined by immunofluorescence, Proc. Soc. Exper. Biol. & Med., 108: 434-439 (1961).

26. COOMBS, R.R.A., DANIEL, M. R., et al., Recognition of the species of origin of cells in culture by mixed agglutination: I. Use of Antisera to Red Cells, Immunology, 4: 55-66 (1961).

27. STAVITSKY, A. B., In vitro studies of the antibody response, Advances Immunol., 1: 211-261 (1961).

28. FISHMAN, M., and ADLER, F. L., Antibody formation initiated in vitro. II. Antibody synthesis in X-irradiated recipients of diffusion chambers containing nucleic acid derived from macrophages incubated with antigen, J. Exper. Med., 117: 595-602 (1963).

29. MOUNTAIN, I. M., Cytopathogenic effect of antiserum to human malignant epithelial cells (strain HeLa) on HeLa cell culture, J. Immunol., 75: 478-484 (1955).

30. GOLDSTEIN, G., and MYRVIK, Q. N., The differentiation of cytotoxic and hemagglutinating antibodies in anti-HeLa cell rabbit sera, J. Immunol., 84: 659-661 (1960).

31. FRANKS, D., GURNER, B. W., et al., Results of tests for the species of origin of cell lines by means of the mixed agglutination reaction, Exper. Cell Res., 28: 608-612 (1962).

32. KELUS, A., GURNER, B. W., et al., Blood group antigens on HeLa cells shown by mixed agglutination, Immunology, 2: 262-267 (1959).

33. HÖGMAN, C. F., Blood group antigens on human cells in tissue culture. Exper. Cell Res., 21: 137-143 (1960).

34. COOMBS, R.R.A., DANIEL, M. R., et al., Recognition of the species of origin of cells in culture by mixed agglutination: II. Use of heterophile (anti-forssman) sera. Internat. Arch. Allergy, 19: 210-226 (1961).

35. MURPHY, W. H., Jr., WIENS, A. L., et al., Impairment of innate resistance by triiodothyronine, Proc. Soc. Exper. Biol. & Med., 99: 213-215 (1958).

36. COX, R. P., and MacLEOD, C. M., Hormonal induction of alkaline phosphatase in human cells in tissue culture, Nature, London, 190: 85-87 (1961).

37. COX, R. P., and MacLEOD, C. M., Alkaline phosphatase content and the effects of prednisolone on mammalian cells in culture, J. Gen. Physiol. 45: 439-485 (1962).

38. SIDMAN, R. L., The direct effect of insulin on organ cultures of brown fat, Anat. Rec., 124: 723-739 (1956).

39. RIVERA, E. M., and BERN, H. A., Influence of insulin on the maintenance and secretory stimulation of mouse mammary tissues by hormones in organ culture, Endocrinology, 69: 340-353 (1961).

40. FELL, H. B., and MELLANBY, E., The effect of l-triiodothyronine on the growth and development of embryonic chick limb-bones in tissue culture, J. Physiol., 133: 89-100 (1956).

41. WESSELLS, N. K., Thyroxine initiation of epidermal differentiation as studied in vitro in chemically defined medium, Exper. Cell. Res., 24: 131-142 (1961).

42. LAWSON, K., The differential growth-response of embryonic chick limb-bone rudiments to triiodothyronine in vitro. III. Hormone concentration. J. Embryol. & Exper. Morphol., 11: 383-398 (1963).

43. GAILLARD, P. J., Observations on the effect of thyroid and parathyroid secretions on explanted mouse radius rudiments, Develop. Biol., 7: 103-116 (1963).

44. KAHN, R. H., Vaginal keratinization in vitro, Ann. New York Acad. Sc., 83: 347-355 (1959).

45. FELL, H. B., The influence of hydrocortisone on the metaplastic action of vitamin A on the epidermis of embryonic chicken skin in organ culture, J. Embryol. & Exper. Morphol., 10:389-409 (1962).

46. NICOLL, C. S., and MEITES, J., Effects of cortisol and thyroxine on prolactin secretion by rat anterior pituitary in vitro, Fed. Proc., 21 (No. 2) 1196 (1962) abstract.

47. ALGARD, F. T., Action of sex hormones on dependent tumors in cell and organ culture systems, Nat. Cancer Inst. Monogr., 11:215-226 (1963).

48. RIVERA, E. M., ELIAS, J. J., et al., Toxic effects of steroid hormones on organ cultures of mouse mammary tumors, with a comment on the occurrence of viral inclusion bodies, J. Nat. Cancer Inst., 31: 671-687 (1963).

49. KULLANDER, S., Studies on the growth and hormone production of rat hypophysis in tissue culture. Acta endocrinol. 43: 147-154 (1963).

50. PETROVIC, A., Fonctions thyréotrope et corticotrope de la préhypophyse en culture organotypique, Biol. Méd., 51: 299-307 (1962).

51. SCHABERG, A., The corticotrophic activity of the anterior hypophysis in vitro, Nat. Cancer Inst. Monogr., 11: 127-141 (1963).

52. GOSPODAROWICZ, D., and LEGAULT-DÉMARE, J., Étude de l'activité biologique in vitro des hormones gonadotropes. IV. Action synergique de la prolactine et de l'hormone chorionique humaine sur le corps jaune de rat in vitro, Act. endocrinol., 42: 509-513 (1963).

53. PAVIĆ, D., The effect of gonadotrophic hormones on young rat ovaries grown in organ culture, J. Endocrinol. 26: 531-538 (1963).

54. LASNITZKI, I., The effect of testosterone propionate on organ cultures of the mouse prostate, J. Endocrinol. 12: 236-240 (1955).

55. PUCK, T. T., and FISHER, H. W., Genetics of somatic mammalian cells. I. Demonstration of the existence of mutants with different growth requirements in a human cancer cell strain (HeLa), J. Exper. Med., 104: 427-434 (1956).

56. MORGAN, J. F., Tissue culture nutrition, Bact. Rev., 22: 20-45 (1958).

57. LEVINTOW, L., and EAGLE, H., Biochemistry of cultured mammalian cells, Ann. Rev. Biochem., 30: 605-640 (1961).

58. DAVIDSON, J. N., (chairman), Nutrition at the cellular level, Proc. Nutrition Soc., 19: 38-68 (1960).

59. FELL, H. B., The application of tissue culture in vitro to embryology, J. Roy. Microscop. Soc., 60: 95-112 (1940).

60. GAILLARD, P. J., Morphogenesis in animal tissue cultures, J. Nat. Cancer Inst., 19: 591-607 (1957).

61. GROBSTEIN, C., Tissue interaction in the morphogenesis of mouse embryonic rudiments in vitro, in Rudnick, D., (ed.), Aspects of Synthesis and Order in Growth, 13th Symposium of The Society for the Study of Development and Growth, Princeton, N.J., Princeton University Press, Chap. 10 (1954).

62. KAHN, R. H., Organ culture in experimental biology, Univ. Michigan M Bull., 24: 242-252 (1958).

63. GROBSTEIN, C., Differentiation of vertebrate cells. in Brachet, J., and Mirsky, A. E., (ed), The Cell, New York Academic Press, Vol. 1, pp. 437-496 (1959).

64. DAWE, C. J., (ed.), Symposium on organ culture: Studies of development, function, and disease, Nat. Cancer Inst. Monogr., 11: 1-80 (1963).

65. FELL, H. B., Techniques of bone cultivation, Meth. Med. Res., 4: 234-237 (1951).

66. FELL, H. B., and MELLANBY, E., The effect of hypervitaminosis A on embryonic limb-bones cultivated in vitro, J. Physiol., 116: 320-349 (1952).

67. BIGGERS, J. D., GWATKIN, R. B. L., et al., Growth of embryonic avian and mammalian tibiae on a relatively simple chemically defined medium, Exper. Cell Res., 25: 41-58 (1961).

68. GROBSTEIN, C., Trans-filter induction of tubules in mouse metanephrogenic mesenchyme, Exper. Cell Res., 10: 424-440 (1956).

69. GROBSTEIN, C., Interactive processes in cytodifferentiation, J. Cell Comp. Physiol., 60 (Suppl. 1): 35-48 (1962).

70. Cancer Chemotherapy National Service Center, Protocols for screening chemical agents and natural products against animal tumors and other biological systems, Cancer Chemotherapy Rep., 25: 1-66 (1962).

71. TOPLIN., I., Experiences with the tissue culture system in large scale cancer chemotherapy screening, Cancer Res., 21: 1042-1046 (1961).

72. McBRIDE, T. J., SCHMITTER, R. W., et al., Screening of synthetic compounds by a tissue culture method, Cancer Res., 23 (No. 2, part 2): 1-7 (1963).

73. TOPLIN, I., A tissue culture cytotoxicity test for large scale cancer chemotherapy screening, Cancer Res., 19: 959-965 (1959).

Chapter V.
TECHNIQUES AND PROCEDURES

MEASUREMENT OF GROWTH

Growth is defined as increase in mass of living substance and is measured usually as increase in cell number, size or both. Often it is sufficient to measure only one of these parameters.

Ideally procedures for growth measurements of cell or organ cultures should have the following characteristics: (a) be applicable at any time without destroying the culture, (b) provide a true measure of increase in mass, and (c) be simple enough to be applied routinely.

A number of methods for measurement of growth have been described. Several, such as cell or nuclear counts, measurement of packed cell volume, or dry weight determinations, are direct methods. Others, such as measurement of nucleic acids, are indirect and depend upon known relationships to cell number or cell mass.

1. ENUMERATION OF CELLS AND NUCLEI

When cells can be grown as monodisperse suspensions or can be dispersed adequately and maintained as a stable suspension, cell counting techniques are preferred. For cells which are difficult to disperse the nuclear count technique is more precise.

A. TOTAL CELL COUNT: HAEMOCYTOMETER

The haemocytometer count is the only feasible means of routine cell count in most laboratories. This method, however, is both time consuming and subject to substantial error. A single count, done carefully, may require 10-15 minutes and even under the best conditions, the method is subject to not less than 10% variation.

1. Clean a haemocytometer and coverglass by first rinsing in water, then in absolute ethanol and finally in acetone. Dry and polish with lens paper. Dry the coverglass with a piece of lint-free cloth.

2. Seat the coverglass firmly on the haemocytometer so that it covers both counting chambers.

3. Dilute 0.5 ml of cell suspension to give a final concentration of approximately 1×10^5 - 2×10^5 cells/ml. If fewer than 100 or more than 200 cells are present in the chamber, the counting error is increased. Use of citric acid-crystal violet solution as a diluent (p. 242) minimizes clumping, stains the nuclei, and thus facilitates counting.

4. With a capillary pipette carefully fill both chambers of the haemocytometer. Be careful not to allow the fluid to overflow. Count the cells in each corner square and the center square of each chamber. The total number of cells counted in the ten squares x 1000 x the dilution previously made equals the number of cells per ml of the original suspension.

B. TOTAL CELL COUNT: ELECTRONIC COUNTER

In an attempt to speed up red blood cell counting procedures and to increase precision, an electronic blood cell counter has been developed and made available commercially*. This instrument is admirably suited to the counting of many animal cells grown in culture (1). The instrument operates on the principle of "electronic gating". A small aperature (100 μ in diameter and 75 μ long) is interposed between two large surface platinum electrodes which are immersed in electrolyte solution (saline). An electric current flows between the electrodes and through the orifice. By applying a vacuum, cells which are suspended in the saline on one side of the orifice are drawn through the orifice. As a cell passes through the orifice, being essentially a non-conductor, it causes a drop in the current which is picked up as an impulse, amplified and recorded on a decade counter. A mercury manometer with appropriately placed electric leads permits metering a sample of exactly 0.5 ml. To eliminate background count the threshold selector should be set so as just to include all cells in the population. A single count, including 1000 or more cells is completed in approximately 20 seconds. A 0.5 ml sample of culture permits 5 or more successive counts on the same sample. The counting error is 1-2%.

1. Measure 24.5 ml of 0.85% NaCl (saline) into a 50 ml beaker. (This may be done quickly and accurately with a "Palo Pipettor"**). Withdraw 0.5 ml of sample to be counted and mix with the saline. This will yield a 1:50 dilution of the original sample.

2. Mix by pouring from one beaker to another 7-8 times.

* Coulter Electronics, Hialeah, Florida.
** Palo Laboratory Supplies, New York 7, N. Y.

3. Place the sample beaker on the platform of the electronic counter with the aperture tube and the electrode immersed in the suspension. The beaker should be positioned so that one side is against the aperture tube. This permits visual observation of the aperture with the microscope mounted on the counter.

4. Open the stopcock to the vacuum source allowing the mercury column to be displaced and the sample to flow through the aperture. Check with the microscope to be certain the aperture is clear.

5. With the appropriate switch, clear the decade counter and then close the stopcock to the vacuum source. As the mercury returns to normal level it will continue pulling sample through the aperture. Appropriately placed "leads" will activate and turn off the counter permitting the counting of exactly 0. 5 ml of sample.

6. Since the original dilution was 1:50 and the amount actually counted was 0. 5 ml the count recorded by the decade counter x 10^2 will equal the number of cells per ml of the original sample.

C. VIABLE CELL COUNT: VITAL STAINS

A number of "vital" staining procedures have been developed in an effort to differentiate between viable and non-viable cells. With some dyes, *e.g.*, methylene blue or various tetrazolium salts, the dye is used to indicate metabolic activity although the relationship of metabolic activity to viability is not at all clear. With others, *e.g.*, eosin Y (2), trypan blue and erythrosin B (3) the ability of viable cells to exclude dye is the criterion used. Serious questions have been raised as to the validity of results obtained with vital stains. At best the results of such methods lack precision and must be interpreted with reservation.

1. Place 0. 5 ml of cell suspension (diluted to contain 1 x 10^5 - 2 x 10^5 cells/ml) in a 12 x 75 mm tube. Add 0. 1 ml of 0. 4% erythrosin B* and mix thoroughly. Allow to stand for 5 minutes but not more than 15 minutes.

2. With a capillary pipette, fill a haemocytometer as for cell counting (p. 155).

3. Make a total cell count and a count of unstained cells. Assuming those cells which are unstained to be viable, express the results as % viable cells.

* 0.4 gms erythrosin B in 100 ml of BSS.

D. VIABLE CELL COUNT: PLATING

Another method for determining the number of viable cells is based on the ability of individual cells to multiply and give rise to colonies which can be enumerated. It gives a relative measure of viable cells. Media and conditions of growth have not been worked out to make the method feasible for many cell cultures. It is also too tedious to be useful as a rapid, routine technique except in isolated instances. The method is described in detail on page 52.

E. ENUMERATION OF CELL NUCLEI

Cells which are resistant to dispersion can be enumerated by the method of nuclear counts. The technique involves treatment with citric acid-crystal violet to destroy the cytoplasm and stain the free nuclei. The cells are then counted using a haemocytometer (4). The additional steps of handling and time involved make the method rather cumbersome for routine counting. Moreover, if a significant number of multinucleate cells are present, an obvious error is introduced.

2. MEASUREMENT OF PACKED CELL VOLUME (PCV)

Growth of cells in suspension culture, or in some cases in monolayer culture, can be measured conveniently by determining the packed cell volume of a given sample of culture (6). Increases in PCV can be used, therefore, to calculate the mean cell volume (6). In measuring PCV it is very important to use a constant gravitational force from experiment to experiment.

A. HOPKINS TUBE*

The Hopkins tube is a graduated tube 16 mm in diameter, with a total capacity of 10 ml and having a capillary stem with a capacity of 0.05 ml calibrated in 0.01 ml divisions. Five or ten ml of sample is placed in the tube which is then centrifuged at 600 G in a bucket type centrifuge for 15 minutes. An angle-head centrifuge will slant the pellet and make it difficult to read accurately. The packed cell mass can be read directly (7). The chief disadvantage of the Hopkins tube is that the graduations are relatively coarse and rather large samples are required to get accurate readings.

B. VAN ALLEN HEMATOCRIT OR THROMBOCYTOCRIT TUBES**

The use of the Van Allen hematocrit tube for measurement of PCV has been described by Waymouth (5). It has the advantage that the tubes are readily available and are easy to use. The exact volume of suspension to be centrifuged is not measured in ml but a graduated portion of the tube is filled and following centrifugation the PCV is read directly as a percentage of the sample. The chief disadvantage of the method is that a rather heavy suspension is required to get a cell pack which can be accurately measured. The thrombocytocrit tube offers some advantage in this regard as it not only has a larger capacity but the stem is narrower and has finer graduations. To use the thrombocytocrit tube a measured volume is filled into the bulb of the tube as with the Hopkins tube and following centrifugation the volume is read directly from the calibrations on the stem. Both the hematocrit and the thrombocytocrit tubes require spring clip closures.

* Kimble Glass Co., Vineland, N. J.

** Eberbach & Son Co., Ann Arbor, Mich.

3. MEASUREMENT OF DRY WEIGHT

A useful device for following the *in vitro* development of some cell populations is change in total cell mass as determined by changes in dry weight. To be meaningful this must be done with cultures in which the cells readily may be freed of medium components so that the dry weight, as determined, is a true measure of the weight of the cells.

Increase in dry weight indicates a net increase in cell mass. Coupled with measurement of packed cell volume and cell number it makes possible a calculation of net synthesis and changes in degree of hydration at both the cell and population levels. The chief limitations to the method are that it is time consuming and that rather large samples are required for accurate determinations. A precaution to be taken is that all samples be dried to a constant weight under specified conditions.

1. Aseptically remove a sample of the cells to be weighed to a conical centrifuge tube. The size of the sample will depend upon the density of the suspension and can vary from 1 to 25 ml. Wash with BSS (not more than three times) and resuspend in 1 ml of BSS. Excessive washing will lead to leakage of materials from the cells.

2. Transfer the sample quantitatively to a previously weighed aluminum planchet and dry to constant weight over P_2O_5. The drying process may be speeded by, (a) prior lyophilization of the specimen, or (b) use of a drying oven in place of P_2O_5.

4. CHEMICAL METHODS

The close relationship of protein and nucleic acid synthesis to cell growth has led many investigators to utilize serial measurement of these components as indicators of growth *in vitro*. Variations in the mean content of protein, RNA and DNA are such that no one measurement can serve as a reliable measure of population development. Use of one or more of these determinations in conjunction with direct methods, as outlined above, provides a much more reliable index of growth. The principle chemical methods employed are given below. Several have been modified or adapted for tissue culture use (8-11).

A. EXTRACTION OF NUCLEIC ACIDS AND PROTEINS

Numerous methods for the extraction of nucleic acids and proteins have been proposed during the past few years. The subject of determination of nucleic acids in biological materials was reviewed in 1961 by Hutchinson and Munro (12) who concluded that the method of choice is a modified Schmidt-Thannhauser procedure (13). Such a method is outlined (p. 162). It should be emphasized that while this method is suitable for analytical purposes, it is not adequate for preparation of biologically active nucleic acids. For such purposes a phenol extraction method usually is employed.

B. METHODS OF ANALYSIS

1. Protein

a. Micro-Kjeldahl Method

Since many proteins have similar nitrogen levels, the micro-Kjeldahl technique often is the method of choice. The technique consists of the conversion of nitrogen to ammonia and titration with a standard acid. Preparation of a standard curve permits conversion of the nitrogen values obtained to protein units (14).

b. Nessler Reaction

Although the micro-Kjeldahl method is extremely accurate, this analysis is much more time consuming than other methods available. The method to be described below is Nesslerization (15). As in the Kjeldahl procedure, the method utilizes the digestion of a sample and conversion to ammonium ions. However, the ammonia is subsequently complexed with a mercury reagent (Nessler's solution) and the endpoint read spectrophotometrically. This method is fairly accurate through the range of 20 to 300 μg of nitrogen (16).

(continued on page 163.)

FLOW SHEET FOR THE SEPARATION OF RNA, DNA AND PROTEIN IN MAMMALIAN CELLS CULTURES *IN VITRO.*

Cell Pellet (0.5 to 2.0 x 10^6 cells)

↓ 1.0 ml 0.5N $HClO_4$
Allow to extract at 4°C for 30 min.
Centrifuge 4°C, 2000 RPM

Supernatant → Acid-soluble phosphate solution

Residue
Wash 1x in absolute C_2H_5OH containing 0.2 N potassium acetate
Centrifuge at 2000 RPM at 4°C

Supernatant →

Residue
Extract 2x with 1 ml portions of 3:1 C_2H_5OH-Ether mixture at 50°C for 30 min.
Centrifuge at 2000 RPM 4°C for 15 min.

Combined Supernatants → Phospholipid fraction

Residue (lipid-free)
Resuspend in 1.0 ml 0.1N KOH 16-18 hrs. at 37°C.

↓

Solubilized cell material

Add 0.034 ml 6N HCl and 1.0 ml 0.5N $HClO_4$
Allow precipitate to form at 4°C for 30 min.
Centrifuge at 2000 RPM 4°C for 15 min.

Supernate → RNA
Mejbaum orcinol reaction

Residue - (DNA & Protein)
Resuspend in 1.0 ml 0.5N $HClO_4$
Heat 90°C for 15 min.
Cool and centrifuge at 2000 RPM 4° C for 15 min.

Supernate → DNA
Burton modification diphenylamine reaction

Residue - (Protein)
Resuspend in 2.0 ml 1N NaOH (may require overnight incubation at room temperature to put protein into solution).

↓

Protein Solution → Protein
Lowry modification Folin-Ciocalteau reaction

MATERIALS

1. Sample of protein (containing about 100 µg nitrogen).
2. Standard ammonium sulfate (containing exactly 100 µg nitrogen).
3. Concentrated sulfuric acid (low in nitrogen), reagent grade.
4. Hydrogen peroxide, 30%, reagent grade.
5. Nessler's reagent*.
6. Digestion tubes, pyrex (Folin-Wu tubes, 20 x 200 mm, graduated at 35 ml and 50 ml).
7. Measuring pipettes, 1 ml.
8. Blunted stirring rods (7 mm x 25 cm)
9. Spectrophotometer
10. Metal digestion racks for Folin-Wu tubes

PROCEDURES

1. Pipette duplicate 1 to 2 ml samples of unknown protein solution, reference standards and blanks (diluent for unknown sample) into Folin-Wu tubes.

2. Add 0.5 ml of concentrated H_2SO_4 to each tube.

3. In a well-ventilated hood, heat tubes over a low burner for 15 min. to 20 min. When water has evaporated, turn flame up and continue to heat. The tubes will release dense fumes of sulfuric acid. Continue to heat for another 5 to 10 minutes.

4. Remove tubes from heat and cool for one minute.

5. Add one drop of H_2O_2 and reheat tubes over a low flame for 10 to 15 minutes. If color persists after this heating, allow tubes to cool, add another drop of H_2O_2 and reheat over a low flame for 5 more minutes. Finally, heat tubes over a higher flame for 10 to 15 minutes.

6. Allow tubes to cool.

7. Add distilled water to the 35 ml mark and place a blunted stirring rod in each tube.

8. With constant stirring, add 5 ml of Nessler's reagent to each tube. Permit reaction to work for 30 minutes.

* Fisher Scientific or Harshaw Chemical Co.

9. Read optical density on spectrophotometer at 440 mμ in order in which Nessler's reagent was added.

10. Using a reference standard containing exactly 100 μg of nitrogen the $\text{mg of protein} = \frac{(\text{OD of unknown - OD of blank})}{(\text{OD of standard - OD of blank})} \times 0.1 \times 6.25$. The figure 6.25 is the "protein factor" and assumes that animal protein contains 16% nitrogen.

c. Lowry Modification of the Folin-Ciocalteau Method

The determination of protein content by this method is based on the color reaction of the aromatic amino acids tyrosine and tryptophane with the Folin-Ciocalteau phenol reagent. The method assumes a constant tyrosine and tryptophane content in protein. The Lowry modification (17) is presented here.

MATERIALS

1. Sample to be tested (in 1N NaOH)
2. Sodium carbonate, reagent grade
3. $CuSO_4 \cdot 5H_2O$, reagent grade
4. Potassium tartrate, reagent grade
5. Folin-Ciocalteau reagent*
6. Phenolphthalein
7. Crystalline bovine serum albumin**
8. Beckman DU spectrophotometer***

PROCEDURE

1. Add 0.5 ml of 1% $CuSO_4 \cdot 5H_2O$ to 0.5 ml of 2% potassium tartrate. Add this mixture, with careful stirring to 50 ml of 2% Na_2CO_3. This is Reagent A.

2. Add 0.5 ml of an appropriately diluted sample to 5 ml of Reagent A. Incubate at 25°C for 10 minutes.

3. Titrate an aliquot of the Folin-Ciocalteau reagent to the phenolphthalein endpoint with 1N NaOH and on the basis of this titration dilute the reagent to 1N with distilled water.

* Eimer and Amend, Fisher Scientific Co., N. Y.
** Armour Laboratories, Chicago, Illinois.
*** Beckman Instruments, Inc., Fullerton, Calif.

4. Rapidly add 0. 5 ml of the 1N Folin reagent to the 5. 5 ml of sample plus Reagent A. Mix uniformly as it is added.

5. At 30 minutes read the blue color which develops in a Beckman DU spectrophotometer at 750 mμ.

6. Construct a standard curve using crystalline bovine serum albumin over the range of 0-300 μg.

7. Read the values of protein of the unknown samples from this curve.

8. For greater accuracy the data can be analyzed by linear correlation method. (18).

2. Ribonucleic Acid

A widely used method for the quantitative determination of ribonucleic acid (RNA) involves measurement of the pentose content by the Orcinol reaction. This depends upon adequate separation from interfering substances. A modification of the Mejbaum reaction (19) is presented here.

MATERIALS

1. Sample to be tested (fractionated by a method which insures separation from interfering substances while preserving the RNA, see p. 162).

2. Purified yeast RNA* in phosphate buffer at pH 6. 8.
3. Ferric ammonium sulfate, reagent grade.
4. Crystalline orcinol (recrystallized from Benzene)
5. Concentrated HCl (12N), reagent grade
6. Beckman DU spectrophotometer**.

PROCEDURE

1. Dissolve 13. 5 grams of ferric ammonium sulfate and 20 grams of recrystallized orcinol in 500 ml of doubled distilled water. Store this as a stock solution in the cold (4°C).

* Nutritional Biochemicals, Inc., Cleveland, Ohio
** Beckman Instruments, Inc., Fullerton, Calif.

2. To prepare a working orcinol reagent add 5 ml of orcinol solution to 85 ml of concentrated HCl and bring to 100 ml with doubled distilled H_2O.

3. Add 1 ml of sample (diluted to contain the equivalent of 4-40 μg of pentose) to 3 ml of the working orcinol reagent and heat in a boiling water bath for 20 minutes.

4. Cool and read the absorption of the green color which develops in a Beckman DU spectrophotometer at 670 mμ.

5. Construct a standard curve from data obtained using quantities of yeast RNA ranging from 0-100 μg/ml of aqueous solution.

6. Read the values of the RNA in the unknown samples from the standard curve.

7. For greater accuracy the data can be analysed by linear correlation methods (18).

3. Deoxyribonucleic Acid

Quantitative determination of deoxyribonucleic acid (DNA) is most commonly done by measurement of deoxyribose by the diphenylamine reaction originally described by Dische and modified by Burton (20).

MATERIALS

1. Sample to be tested (extracted by a method which will insure separation from interfering substances while preserving the DNA, see p. 162).
2. Diphenylamine (twice recrystallized from 70% ethanol.)
3. Redistilled glacial acetic acid, reagent grade.
4. Concentrated H_2SO_4, reagent grade.
5 Acetaldehyde, reagent grade
6. Crystalline herring sperm DNA*
7. Beckman DU Spectrophotometer**

* Nutritional Biochemicals, Inc., Cleveland, Ohio.
** Beckman Instruments, Inc., Fullerton, Calif.

PROCEDURE

1. Diphenylamine stock solution is prepared by adding 1.5 grams of diphenylamine and 1.5 ml of concentrated H_2SO_4 to 100 ml of glacial acetic acid. This stock may be stored at 4°C for several weeks.

2. Just before use add 0.1 ml of acetaldehyde to 20 ml of the diphenylamine stock solution.

3. Add 1 ml of sample (containing 50-500 µg of DNA) to 2 ml of diphenylamine reagent and incubate at 37°C for 14-18 hours). The incubation time should be constant from experiment to experiment!

4. Read the blue color which develops in a Beckman DU spectrophotometer at 600 mµ.

5. Using 0-100 µg of herring sperm DNA in aqueous solution construct a standard curve. Read the DNA values of the unknown samples from this curve.

6. For greater accuracy the data can be analysed by linear correlation methods (18).

5. MITOTIC COEFFICIENT

In many instances, particularly with explant cultures, it is not possible to make serial cell number measurements for following growth of cultures. In such cases it may be helpful to determine the mitotic coefficient. Practically, this is done by determining the number of mitoses per 100 cells. Such a method is dependent upon the ability to examine the culture with phase contrast optics at high magnification, or the culture alternatively must be fixed and stained. A comparison of the mitotic coefficients of cultures under varying experimental conditions often is useful. Where the mitotic rate is low, prior treatment with colchicine may be necessary (p. 198). This procedure will yield the proportion of cells in a population undergoing division in a given unit of time and is known as the mitotic index (see glossary).

6. RADIOAUTOGRAPHY

With increased availability of labeled compounds and with simplification of procedures (21, 22), the technique of radioautography has become a valuable and critical tool for the study of cellular metabolism and for the localization of inter- and intracellular organelles. Cell cultures provide an ideal starting material for investigating the synthesis and the localization of proteins and nucleic acids (23-25) and for the effect of various agents (virus, irradiation, *etc.*) on cell metabolism (24, 26).

The following provides a general description of the method. The choice of isotope, compounds, cells and media will depend upon the needs of the investigator.

MATERIALS

1. Stock monolayer culture of cells on coverglasses.
2. Sterile screw-cap 16 x 125 mm tubes each containing one 9 x 22 mm #1 coverglass.
3. Sterile cotton plugged 1 ml serological pipettes.
4. Sterile media (p. 220) containing isotope (*e. g.* 1μc/ml tritiated cytidine-uridine; 1μc/ml tritiated amino acid; 0. 3 μc/ml tritiated thymidine).
5. Sterile fine forceps.
6. Sterile balanced salt solution (BSS) (p. 214).
7. Columbia staining jars.
8. Glacial acetic acid: 95% ethanol (1:3) fixative.
9. Clean microscope slides (1 x 3 inches).
10. NTB3 Liquid Emulsion*.
11. Water bath regulated at 42-45°C.
12. Dark room.
13. Wratten series #1 Safelight filter with 15 watt bulb.
14. Coplin jars.
15. Light-tight black Bakelite microscope slide box.
16. D-11 developer*.
17. Acid fixer with hardener*.
18. 0. 25% Toluidine Blue stain (pH 6).
19. Canada balsam or Euparol.

* Eastman-Kodak, Rochester, N.Y.

PROCEDURE

1. A monolayer coverglass culture in mid-logarithmic growth phase should be used.

2. Decant off the medium and add one ml of fresh medium containing isotope.

3. After various periods (minutes or hours) of incubation at 35°C, remove the coverglasses from tubes with fine forceps. Carefully rinse each coverglass in four changes of warm BSS and fix for forty minutes in the acetic acid: alcohol fixative. Remove the cover-glasses from fixative and allow to air dry. For pulse labeling, the isotope-containing medium may be rinsed off with fresh medium and the culture allowed to continue for varying periods of time.

4. Mount the coverglasses for each time period on a clean microscope slide with the cells facing up. Use a minimum of Euparol as the mounting medium. Keep the coverglasses toward one end of the slide. [Where controls are required, such as ribonuclease, cut the coverglasses in half by lightly scratching the surface with a diamond pointed pencil and then break it into two equal halves. Incubate one half of each control coverglass in ribonuclease* (1 mg/ml distilled water, pH 6. 5-7. 0) for one hour at room temperature. Rinse in water and allow to air dry.]

5. With a diamond pencil, label opposite side of slide with name and time of incubation and place it in a drying oven (45°C) for 24 hours.

6. Coat the slide with NTB3 liquid emulsion in the dark room. A sample of the stock emulsion is melted by incubating the bottle in a 42-45°C water bath for approximately 45 minutes. Check melted emulsion for bubbles and scoop out any that stand on surface. The dry slides should be arranged back to back in a coplin jar and then the jar placed in the water bath to pre-warm the slides. The dry slides, two at a time, are dipped once into the melted emulsion for 4-5 seconds. Withdraw slides and allow to drain in a vertical position for a moment over the emulsion. Separate the slides, wipe the back of the slide with a piece of tissue paper and dry in a upright position. (This takes about one and a half hours. The safelight should be off during the drying period)

7. When slides are dry, transfer to a light-tight slide box, seal with black tape and leave box standing on edge (slides exposed horizontally) at room temperature for at least 48 hours. Periods up to six months may be necessary.

* California Corp. for Biochemical Research, Los Angeles, California.

8. Develop the slides in the dark room as follows:

   a. two minutes in D-11 developer
   b. rinse in water
   c. fix in acid fixer for 2-5 minutes
   d. rinse in running water for 20 minutes (lights may be on now)
   e. rinse briefly in distilled water
   f. allow to dry and store for future staining

9. Stain slides in 0.25% toluidine blue, rinse in 95% ethanol and air dry. Slides may be sealed with a coverglass using Canada balsam or Euparol as mounting medium.

The interpretation of the radioautography will depend on the isotope used, the time of incubation and the method of analysis. For example, if tritiated thymidine is used to investigate DNA synthesis, one may estimate the premitotic non-synthetic period (G_2) by determining the time required for labeled DNA to show up in metaphase chromosomes. The grain count over metaphase cells will increase with time after the addition of label, being zero until the first cell passes through the G_2 period and then will increase in number for a period of time after which there will be no further increase. The time required from the first signs of grains to where no further increase occurs will approximate the DNA synthetic period (S).

The length of time which cells spend in mitosis (M period) may be ascertained by the following formula

$$t_{mitosis} = \frac{MT}{0.693}$$

where M x 100 is the mitotic index and T is the generation time, assuming the culture is completely asynchronous (*i.e.* mitotic index is constant with time). The G_1 can be then determined knowing that generation time is the sum of G_1, S, G_2 and M periods.

7. INCREASE IN AREA

In following the growth of explant cultures or of some types of organ cultures, a useful measurement is the increase in surface area. This is usually determined as the increase in surface area of a plane surface. It is of limited value since no consideration is given to increase in volume.

A. CAMERA LUCIDA AND PLANIMETER

A camera lucida* is a device which fits over the ocular of a monocular microscope and projects the image from the ocular onto a plane surface where the outline may be traced. The planimeter is a device which measures the area of any plane surface by moving a pointer around its boundary and reading the indicator on a scale.

B. OCULAR MICROMETER

A more rapid and convenient method, though less precise, is the use of a calibrated ocular micrometer. The diameter of a colony or explant may be measured in 2 or more directions in one plane. Taking the mean value of these measurements, one can calculate roughly surface area from the formula for the area of a circle.

* E. Leitz

8. MEASUREMENT OF CELL SIZE

Direct measurement of cell size can be accomplished by two techniques. A calibrated ocular micrometer can be used to measure mean cell diameter from which cell volume can be calculated. It involves the basic assumption that the cells, when free in suspension, are spherical. The "Coulter Electronic Cell Counter and Cell Size Analyzer"* gives a direct measure of volume. An indirect measure of mean cell volume can be obtained by dividing packed cell volume by total cell number (p. 159).

In any measure of cell size it is important to appreciate that suspending fluids, washing solutions, *etc.*, may cause changes in cell size because of osmotic phenomena. It is desirable, whenever possible, to suspend the cells in the growth medium for measurement. If other suspending solutions are used appropriate controls are necessary.

A. OCULAR MICROMETER

1. Transfer a sample of the cells to a 12 x 75 mm tube. Dilute the cells, in growth medium, to give a concentration of approximately 5×10^5 cells/ml.

2. Place a drop of the cell suspension on a clean microslide and cover with a clean coverglass. Do not compress!

3. Bring the cells into focus with the 40 X objective and measure the diameter of each of 100 cells using a calibrated ocular micrometer.

4. Applying the formula $\frac{4}{3}\pi r^3$, calculate the mean volume of the cells.

B. ELECTRONIC COUNTER

1. Measure 49 ml of growth medium or saline into a 50 ml beaker. Add 1 ml of cell suspension. (Saline may be used providing significant changes in cell size do not occur when the cells are transferred from growth medium.)

2. Pour the sample from one 50 ml beaker to another 7-8 times to obtain a uniform distribution of cells in the sample.

3. Place the beaker on the platform of the Coulter Counter and make a total cell count at a threshold level which will include all of the cells in the sample (p. 156).

* Coulter Electronics, Hialeah, Florida.

4. Raise the threshold 5 or 10 divisions and make another count. Repeat at 5-10 division intervals. Stir carefully between counts to avoid settling of cells.

5. Determine the differences in count between each two successive threshold settings. This represents the increment of the population falling in this particular size range. Plot these values against threshold settings. Data can be also plotted as percentage of the total count as a function of the increments between successive threshold settings.

6. To obtain absolute values for cell size the counter must be calibrated with particles of known volume, *e. g.*, erythrocyte, polystyrene particles*.

C. MEAN CELL VOLUME

Measure packed cell volume (PCV) as directed on page 159. Prior to centrifuging the sample make a total cell count (p. 156). Divide the PCV by the total cells (cells/ml x volume) to obtain a measure of mean cell volume.

* Dow Chemical Company, Midland, Michigan.

CYTOLOGY

Although it is unreasonable to classify cells grown *in vitro* on a morphological basis, for the present this is the most practical means of doing so. The appearance of a cell depends to a large extent upon the method of culture, the length of time in culture, the medium in which it is growing, the gaseous environment, *etc.* With few exceptions, (*e.g.*, nervous tissue, muscle, *etc.*) one can classify conveniently all cells cultivated as a monolayer into one of three general morphological types, namely (a) fibroblastic (fibroblast-like, mechanocyte), (b) epithelioid or (c) ameboid. In reality, each of these morphological types probably represents an infinite number of potentially different cells given the appropriate environment.

The group of fibroblastic cells are so named because of their similarity to the fibroblast or fibrocyte seen *in vivo.* Such a term is misleading in that these fibroblastic cells may have little in common functionally, chemically or embryologically with the analogous cell *in vivo.* This is demonstrated clearly by the simple comparison of a clonal fibroblastic cell seen as a monolayer when contrasted with the same cell grown in suspension. The fibroblastic cell is identified morphologically in a monolayer on the basis of its spindle or branching shape and growth in a loose network. Its cytoplasm is relatively clear but as shown with phase optics contains a juxta-nuclear area which is known to consist of the Golgi apparatus and vacuoles. The mitochondria generally are observed as filamentous bodies. The large, oval, centrally placed nucleus appears translucent with dust-like chromatin and contains one or more prominent nucleoli.

The epithelioid type of cell characteristically grows as a closely connected sheet of polygonal cells either as a membrane leaving little intercellular space or in a tubular or cyst-like fashion. Generally speaking there is a higher ratio of cytoplasm to nucleus in this type of cell than in either the fibroblastic or ameboid cell. Because of the close apposition of cells, the borders of the culture generally appear smoother in contour.

The amoeboid cells, on the other hand, behave as independent, freely moving units and grow in a rather scattered fashion until the culture becomes crowded. High population densities give rise to monolayers. There is evidence to suggest that these cells may assume the fibroblastic or the epithelioid shape depending on the technique of cultivation. The cells are irregularly shaped and demonstrate rather striking undulating pseudopodia.

FIXATION AND STAINING PROCEDURES

1. METHODS OF FIXATION

The primary purpose of fixation is to alter living cells so as to prevent autolysis, distortion due to subsequent treatment or bacterial decay and at the same time to stabilize their intrinsic chemical and morphological constituents. A number of fixatives are available for general histological use, many of which may be adopted for use in tissue culture. Space does not permit a complete listing nor discussion of the attributes of each (27-29).

The choice of fixatives is dictated by what one wishes to demonstrate. For example, acetic acid should not be used for preparations designed to demonstrate mitochondria. Regaud fixative is suitable for this purpose. Lipids can be preserved only by aqueous fixation and subsequent treatment with appropriate solvents. In addition, the requirements for fixation depend to a large extent on cells or tissues involved. For example, monolayers do not require paraffin embedding and sectioning and are therefore less demanding than explants and fragments which subsequently must be sectioned. In addition, the length of time for fixation also will vary with the tissue. It is for these reasons that the fixation times indicated below are in broad terms.

Cultures grown in plasma clots or other natural media should be washed with several changes of warm BSS prior to fixation to avoid artefacts due to precipitation of medium components.

A. 10% NEUTRAL FORMALIN (4% Formaldehyde)

This is a good fixative for whole mounts such as monolayers or frozen sections but is not recommended for use preceding paraffin embedding. It yields a homogeneous protein distribution and preserves mitochondria and lipids generally. It is not recommended to preserve carbohydrates. Any cytoplasmic basophilia will be accentuated. Fix for 10-60 minutes.

0.85% NaCl*	900 ml	Neutralize formalin by adding an excess of $MgCO_3$ and then dilute in saline.
Formalin (40% formaldehyde)	100 ml	

* Isotonic saline is preferred to distilled water.

B. NEUTRAL BUFFERED FORMALIN (NBF)

This fixative is similar to 10% neutral formalin but avoids a pH shift upon use.

Formalin (40% formaldehyde)	100 ml	Neutralize formalin with an excess of $MgCO_3$ and dilute ingredients with 0.85% NaCl* to one liter.
Na_2HPO_4 (anhydrous)	7.83 gm	
NaH_2PO_4 (anhydrous)	4.21 gm	

C. OSMIUM TETROXIDE

Osmium tetroxide is an excellent fixative for whole mounts such as monolayers but is not recommended for use prior to paraffin embedding and sectioning. It preserves proteins and lipids in a homogenous distribution but is not recommended for carbohydrates. This fixative penetrates slowly so the tissue being fixed must be relatively thin and the time of fixation proportionately longer; usually 2 hours will be maximum and 30-60 minutes will be optimal.

Osmium tetroxide (OsO_4)	1 gm	Using a fume hood, dissolve osmium in water with extreme care as the vapor is highly toxic.
Distilled water	100 ml	

D. FORMALIN-ACETIC ACID-ALCOHOL (FAA)

FAA is recommended for monolayer cultures or for explants to be paraffin embedded and sectioned. It is a good fixative which preserves most proteins. Fix for at least 30-40 minutes.

80% ethanol	90 ml	Neutralize formalin with an excess of $MgCO_3$. This fixative is best made up fresh.
Glacial acetic acid	5 ml	
Neutralized formalin (40% formaldehyde)	5 ml	

* Isotonic saline is preferred to distilled water.

E. CARNOY FIXATIVE

Carnoy fluid is a good non-aqueous fixative to precede specific stains. It is recommended especially for nuclear detail and for staining polysaccharides but is not suitable for delicate or poorly penetrable structures. Fix for 15-60 minutes.

Absolute ethanol	60 ml
Glacial acetic acid	10 ml
Chloroform (reagent)	30 ml

F. BOUIN FLUID

Bouin fluid is recommended for general histology and may be used on monolayers. For explants to be paraffin embedded and sectioned, fix for 2-16 hours, wash in 50% ethanol and then in several changes of 70% ethanol to remove traces of picric acid. Such explants may be stored in 70% ethanol for future dehydration and paraffin infiltration. Monolayers should be fixed for 1/2 - 2 hours, washed in 50% and 70% ethanol as above. Further treatment will depend upon the stain to be used.

Picric acid (1.4 gm/100 ml dist. H_2O)	75 ml
Formalin (40% formaldehyde)	25 ml
Glacial acetic acid	5 ml

G. REGAUD FLUID

This fixative is recommended for explants to be paraffin embedded and sectioned and for demonstrating mitochondria. Fix for 4 days, changing fixative daily and then treat with 3% potassium bichromate for 7 days, changing every second day (post-chromatin). Wash in running water 24 hours, dehydrate, clear, double embed in paraffin and cut sections at 3 μ.

Potassium bichromate, 3% aqueous	80 ml	Neutralize formalin with an excess of $MgCO_3$ and then add chromate.
Neutral formalin (40% formaldehyde)	20 ml	

2. COLLODION STRIPPING TECHNIQUE

The examination of monolayers cultivated in large culture vessels (*e. g.* French square or prescription bottles) is extremely difficult due to the poor optical qualities of these vessels. Furthermore, such cells cannot be removed readily for subsequent fixation and staining. Enders and Peebles (30) described a collodion stripping technique subsequently refined by Reissig, Howes and Melnick (31) which permits the removal of cells from large glass surfaces.

MATERIALS

1. Collodion, U. S. P. *
2. Oil of Cloves

PROCEDURE

1. Rinse vessel with warm BSS.

2. Add Bouin fluid and fix for at least 30 minutes (maximum 2 hours).

3. Dehydrate with two changes each of 70% ethanol, 95% ethanol and absolute ethanol (2 minutes for each change).

4. Dehydrate with anhydrous ether-absolute ethanol (1:1) for 5 minutes.

5. Cover surface of the vessel carrying cells with collodion for at least two hours in unstoppered bottle.

6. After two hours, cap vessel and leave collodion in contact with cells overnight.

7. Pour any remaining collodion back in the stock bottle and let vessel drain evenly by inverting and rotating it. Blow into vessel until only faint odor of ether remains.

8. Run water down opposite side of vessel from tissue. (Do not leave water in vessel more than 4 or 5 minutes).

9. While in contact with water, separate collodion film from glass with forceps or metal spatula.

10. Pour out water and remove film.

* Merck and Co., Inc., Rahway, N. J.

11. Trim that part of film containing cells and cut into 20 mm squares. Press each square onto paper. Transfer squares to 1X3 microslide with cells down.

12. Cover slide with oil of cloves for one to four hours until film is cleared.

13. Wash slide in several changes of absolute ethanol.

14. Transfer to anhydrous ether-absolute ethanol (1:1) for several hours to remove collodion.

15. Store in 70% ethanol until ready to stain.

16. Stain as desired.

3. METHODS OF STAINING

A number of good staining procedures are available. Each procedure has its attributes and many are designed for a specific purpose. Space permits a listing of but a few of the available methods (27-29, 32-35). It is assumed that the reader is familiar with the general techniques for embedding, sectioning and mounting (27-29). It should be recognized that the techniques provided here may require modification in time of fixation, staining and destaining (differentiating) depending on the tissue source, size of tissue and procedure.

A. WRIGHT STAIN

This stain is suggested for cells grown in suspension culture. This procedure is a rapid method for evaluating mitotic index and general morphology of the isolated cell.

MATERIALS

1. Wright stain* 0.1 gm Dissolve in 60 ml of absolute methanol with the aid of a mortar and pestle.

PROCEDURE

1. Centrifuge a 1-5 ml sample of a cell suspension and decant supernatant fluid.

2. Resuspend in a small drop of BSS and make a film of cells on a slide. Spread thinly as for a blood film and allow to dry in air.

3. Cover preparation completely with Wright stain for 1 minute.

4. Add to the Wright stain an equal quantity of distilled water and counterstain for 2-3 minutes.

5. Wash in running water until clear of any excess stain.

6. A permanent preparation can be made by dehydrating slide through a series of ethanols (50%, 80%, 95%, 100%) and clearing in xylene. Then mount by placing two drops of balsam or HSR** on the slide and cover with a coverglass.

* Wright stain -- Allied Chemical and Dye Corp.

** Hartman-Leddon Co., Philadelphia, Pa.

B. MAY-GRÜNWALD-GIEMSA STAIN

This stain is recommended for monolayer cultures to demonstrate differentially ribo- and deoxyribo-nucleoproteins (RNA-Protein and DNA-Protein). DNA-Protein stains red-purple while RNA-Protein stains blue. Ribonuclease treatment for one hour may be used as a control.

MATERIALS

Stock May-Grünwald stain*	2.5 gm	Dissolve in absolute methanol to 1000 ml. Age 1 month. Filter.
Stock Giemsa*	1.0 gm	Dissolve in 66 ml of glycerol at 55-60° C for 1.5-2.0 hrs.; add 66 ml of absolute methanol.

PROCEDURE

1. Wash in three rinses of warm BSS (p. 214).
2. Fix for 5 minutes in absolute methanol. Agitate during fixation.
3. Stain for 10 minutes in filtered stock May-Grünwald solution.
4. Stain for 20 minutes in dilute Giemsa solution (dilute 1:15 in distilled water just before use.)
5. Rinse rapidly in distilled water. (10-20 sec.)
6. Quickly rinse in 2 changes of acetone to dehydrate tissue. Do not let coverslip or slide dry.
7. Clear by rinsing three times in acetone-xylol (2:1), three times in acetone-xylol (1:2) and 10 minutes in fresh xylol.
8. Mount in balsam or HSR.

* Dry powder, certified, Allied Chemical Corp.

C. METHYL GREEN - PYRONIN

MATERIALS

1. Methyl green-Pyronin stain

Methyl green*	0.5%	Dissolve 1 gram methyl green in 200 ml of Walpole buffer. Wash dye in separatory funnel with chloroform until all violet color has been removed from stain. Let stand two days until chloroform has evaporated. Add Pyronin to a final concentration of 0.1%. Store in refrigerator and bring to room temperature before use.
Pyronin**	0.1%	

2. Walpole buffer

0.2 M Glacial acetic acid (HAC)	Add 1.15 ml Glacial acetic acid to 98.85 ml water.
0.2 M Sodium acetate (NaAC)	Add 2.72 gm Sodium acetate to 100 ml of water.
	Mix 150 ml of 0.2 M HAC and 50 ml of 0.2 M NaAC. pH=4.16.

PROCEDURE

1. Fix in cold FAA (p. 177) for 10 minutes (2 hrs. for paraffin embedded tissues).
2. Rinse in 70% and then 50% ethanol.
3. Rinse in distilled water.
4. Stain for 5 minutes in methyl green-pyronin stain.
5. Rinse in distilled water.
6. Dehydrate with tertiary butyl alcohol-ethanol (3:1) 3X.
7. Clear in two changes of xylene (5 minutes each).
8. Mount in balsam or HSR.

* Methyl Green (C.I. 42590) Allied Chemical and Dye Corp.
** Pyronin (C.I. 45005), Ibid.

D. METHYLENE BLUE

This procedure is designed for demonstration of nucleoproteins and mitotic figures.

MATERIALS

1. Methylene blue stain

Methylene blue 1% aqueous solution	10 ml
Citric acid - phosphate buffer pH 5.6 (84 ml of M/10 citric acid; 116 ml of M/5 disodium phosphate (anhydrous)	10 ml
Acetone	25 ml
Distilled H_2O.	140 ml

2. Rosin 2% solution in 95% ethanol

PROCEDURE

1. Fix in FAA (p. 177) for approximately one hour.

2. Rinse in 50% ethanol.

3. Rinse in distilled H_2O.

4. Stain for 15 minutes.

5. Destain in 95% ethanol plus 2% rosin.

6. Dehydrate through absolute ethanol and xylene.

7. Mount in balsam.

   Control slides should be incubated for one hour in 0.01% ribonuclease at 60° C.

E. HEMATOXYLIN AND EOSIN

This procedure is used routinely for explants which have been embedded and sectioned and is recommended for general histology. Nuclei stain blue, while cytoplasm stains pink.

MATERIALS

1. Alum Hematoxylin

Hematoxylin	0.5 gm	Dissolve solids in 70 ml distilled water. Add glycerol and acetic acid; if residue remains, filter; solution is then ready for use.
Ammonium alum: $[(NH_4)_2SO_4 \cdot Al_2(SO_4)_3 \cdot 24H_2O]$	5.0 gm	
$NaIO_3$	0.1 gm	
Glycerol	30.0 ml	
Acetic acid (glacial)	2.0 ml	

2. Bicarbonate

$NaHCO_3$	1.0 gm	Dissolve in 100 ml distilled H_2O.

3. Eosin Y*

Dry powder	0.5 gm	Dissolve in 100 ml H_2O.

PROCEDURE

1. Rinse three times in warm BSS.
2. Fix in neutral buffered formalin for 30 minutes. (If explants, paraffin embed, section, mount and hydrate).
3. Rinse once in distilled water.
4. Stain for 10 minutes in alum-hematoxylin diluted 1:20 in distilled water.
5. Rinse thoroughly in tap water.
6. Expose until cells assume a blue color in 1% aqueous NaHCO .
7. Stain in 0.5% aqueous eosin for one minute.
8. Rinse once in distilled water.
9. Dehydrate rapidly through two changes of acetone to prevent drying.
10. Rinse 3 times in acetone:xylol 2:1.
11. Rinse 3 times in acetone:xylol 1:2.
12. Rinse 3 times in xylol.
13. Clear in xylol (fresh) for 10 minutes.
14. Mount with balsam.

* Allied Chemical & Dye Corp. (C.I. 45380).

F. REGAUD IRON HEMATOXYLIN-MASSON TRICHROME METHOD

This procedure is recommended to stain sectioned tissue such as explants or organ cultures. This combination of dyes was designed as a differential stain for collagenous connective tissue and for reticulum. It is also a good stain for general morphology. Fibrous connective tissue and hyaline cartilage are deep blue; muscle, myelin and red blood cells are various shapes of yellow; cytoplasm pink with deep purple nuclei; fibrin is deep pink and bone matrix is bright red.

MATERIALS

1. Picric acid-alcohol solution - 95% ethanol saturated with picric acid

2. Masson A - 0.3 gm acid fuchsin*
   1 ml glacial acetic acid
   0.7 gm ponceau de xylidine**
   100 ml distilled water

3. Masson B - 1.0 gm phosphomolybdic acid
   100 ml distilled water

4. Masson C - 2.0 ml glacial acetic acid
   100 ml distilled water
   2 - 3 gm aniline blue***

   or

   Light Green - 2 gm Light Green****
   S.F. Yellowish
   100 ml 1% acetic acid

5. Regaud Hematoxylin - 1.0 gm hematoxylin*****
   10 ml absolute ethanol
   10 ml glycerine
   80 ml distilled water
   Allow to age three weeks

6. Eosin-Alcohol - Eosin Y****** in 80% ethanol - 1% solution

7. Iron Alum, (ferric ammonium sulfate), 5% solution

* Allied Chemical & Dye Corp. (C.I. 42685)
** Ibid (C.I. 16150)
*** Ibid (C.I. 42780)
**** Ibid (C.I. 42095)
***** Ibid (C.I. 75290)
****** Ibid (C.I. 45380)

PROCEDURE

1. Fix in FAA (p. 177) for 24 hours.
2. Dehydrate by replacing with 80% ethanol for minimum of 4 hours (may be stored here).
3. Add 5 drops eosin-alcohol to 80% ethanol and allow to stand for $\frac{1}{2}$ hour (to stain small explants so they may be seen in the paraffin blocks).
4. Dehydrate in 95% ethanol for $1\frac{1}{2}$ hours.
5. Dehydrate in absolute ethanol for $1\frac{1}{2}$ hours.
6. Store in cedarwood oil overnight.
7. Rinse in xylene (or benzene) 2 changes, 15 minutes each.
8. Embed at 60° C in a vacuum oven in at least 2 changes of paraffin, $\frac{1}{2}$ hour each or 1 hour each in an oven at atmospheric pressure.
9. Block in paraffin.
10. Section and mount on slides.
11. Place in xylol for 3 minutes.
12. Rehydrate by consecutive changes from absolute ethanol to distilled water, 3 minutes each change (100%, 95%, 70%, 50%, H_2O).
13. Mordant in 5% iron alum for 4 hours at room temperature or 5 minutes at 50° C.
14. Rinse in distilled water.
15. Stain in Regaud hematoxylin for 2-3 minutes at 45-50° C.
16. Rinse in 95% ethanol.
17. Destain in picric acid-alcohol and check with microscope. (Nuclear detail should be visible.)
18. Wash in running water for 10 minutes.
19. Stain in Masson A for 5 minutes.
20. Rinse in distilled water.
21. Destain in Masson B for 3-15 minutes.
22. Stain in Masson C for 2-5 minutes or Light Green for 3 minutes.
23. Rinse in distilled water.
24. Destain in Masson B for 5 minutes.
25. Destain in 1% acetic acid for 2-5 minutes.
26. Dehydrate in 2 changes each of 95% and absolute ethanol.
27. Clear in xylene and mount.

4. CYTOCHEMICAL PROCEDURES

A. PERIODIC ACID-SCHIFF (36) and METHYLENE BLUE

The PAS technique is recommended to demonstrate neutral mucopolysaccharides, glycoprotein and mucoproteins and for glycogen when coupled with amylase digestion. The counterstain (methylene blue) demonstrates cellular basophilia (RNA, DNA, acid mucoploysaccharides). It is designed for sectioned material and is adaptable for monolayers as well.

MATERIALS

1. Periodic acid solution

Periodic acid	1.2 gm	Dissolve periodic acid in 30 ml distilled water, add the sodium acetate and then the ethanol.
Sodium acetate M/5	15 ml	
Absolute ethanol	105 ml	
Distilled water	30 ml	

2. Reducing rinse

Distilled H_2O	60 ml	Dissolve solids in 60 ml distilled water, stir in the ethanol and then add the HCl.
Potassium iodide	3 gm	
Sodium thiosulfate ($Na_2S_2O_3 \cdot 5\ H_2O$)	3 gm	
Absolute ethanol	90 ml	
HCl, 2 N	1.5 ml	

3. Sulfite wash water

10% sodium metabisulfite ($Na_2S_2O_5$)	30 ml	Add the metabisulfite to 540 ml distilled water and then add the HCl.
1 N HCl	30 ml	
Distilled water	540 ml	

4. Schiff Reagent

Basic fuchsin*	2.0 gm	Dissolve stain and metabisulfite in acid. Shake at intervals for 2 hrs. Solution should be straw-colored. Add 500 mg fresh activated animal (continued)
Potassium metabisulfite ($K_2S_2O_5$)	3.8 gm	
0.15 N HCl	200 ml	
Activated charcoal	500 mg	

* Allied Chemical and Dye Corp. (C.I. 42500)

4. (continued) charcoal (to decolorize) and filter. Store in small, dark glass bottles in refrigerator.

5. Methylene blue stain (p. 184).
6. Amylase solution, 1% in distilled water.
7. Rosin, 2% in 95% ethanol.

PROCEDURE

1. Fix in FAA (or Carnoy) (p. 177-178).
2. Paraffin embed and section. (If monolayer, skip to step 4).
3. Mount and hydrate (xylene, 100%, 95% ethanol).
4. Three rinses of 70% ethanol.
5. Place in periodic acid solution for 10 minutes.
6. Rinse thoroughly in 70% ethanol (Do not let stand.)
7. Place in reducing rinse for 5 minutes exactly.
8. Rinse thoroughly in 70% ethanol (Do not let stand.)
9. Treat with Schiff reagent for 45 minutes.
10. Rinse in three changes of sulfite wash water, 2 minutes each, exactly.
11. Wash under running water 10 minutes, rinse in distilled water.
12. Stain in methylene blue solution for 2 minutes.
13. Rinse in distilled water.
14. Destain in 2% rosin solution. (Tissue should be pale blue and the blue stain should not come out in clouds.)
15. Dehydrate through two changes absolute ethanol.
16. Pass through four changes in xylol.
17. Mount in balsam or HSR.

To remove glycogen, hydrate through 70% ethanol to water after step 3 and place in unbuffered 1% amylase solution for 15 minutes at 37°C. Rinse thoroughly in distilled water, then in 50% ethanol for 5 minutes and continue with step 4.

B. PERIODIC ACID-SCHIFF AND ALCIAN BLUE (37).

Neutral and acidic mucopolysaccharides may be simultaneously demonstrated with PAS and alcian Blue. The latter dye stains complex carbohydrates rich in free acidic groups.

MATERIALS

1. Alcian Blue solution

Alcian Blue 8 GX*	0.1 gm	Add acetic acid to water and dissolve dye in the weak acetic acid solution. Filter and add crystal of thymol to preserve.
Glacial acetic acid	3.0 ml	
Distilled water	97.0 ml	

2. Periodic acid solution (p. 188).
3. Reducing rinse (p. 188).
4. Sulfite wash water (p. 188).
5. Schiff Reagent (p. 188).
6. Alum Hematoxylin (p. 185).
7. Neutral buffered formalin (p. 177).
8. 3% acetic acid.

PROCEDURE

1. Fix in neutral buffered formalin.
2. Paraffin embed and section. (if monolayer, skip to step 4)
3. Mount and hydrate (xylene, 100%, 95%, 70%, 50% ethanol, water).
4. Rinse in 3% acetic acid for 3 minutes.
5. Stain 2 hours in Alcian Blue solution.
6. Rinse in tap water and then in 3% acetic acid for 3 - 5 minutes.
7. Wash for 2 minutes in running tap water and then briefly in distilled water.
8. Follow steps 5-11 under Periodic Acid Schiff and Methylene blue (p. 189).
9. Stain 5 minutes in Alum Hematoxylin.
10. Wash 2 minutes in running tap water.
11. Dehydrate through 70%, 95% and 2 changes absolute ethanol followed by 2 changes of xylol.
12. Mount in HSR or balsam.

* Imperial Chemical Industries Ltd., London, England.

C. FEULGEN TECHNIQUE FOR DNA

This procedure is perhaps the most specific staining technique available for demonstrating deoxyribonucleoprotein.

MATERIALS

1. HCl, 1 N
2. Schiff reagent (p. 188).
3. Sulfite wash (p. 188).
4. Fast Green FCF*, 0.01% in 95% ethanol

PROCEDURE

1. Fix in Carnoy fixative for 20-30 minutes (p. 178).
2. Rinse in absolute ethanol for 3-5 minutes.
3. Hydrate through 95% and 70% ethanol.
4. Rinse in distilled water.
5. Incubate in 1 N HCl at 60°C for 10 minutes.
6. Rinse in distilled water.
7. Stain in Schiff reagent for 10 minutes.
8. Rinse in three changes of sulfite wash, two minutes each.
9. Wash in running water for 15 minutes or until sulfite smell is gone.
10. Rinse in 70% ethanol.
11. Counterstain in Fast Green FCF solution for 5 seconds.
12. Dehydrate in absolute ethanol and clear in xylene.
13. Mount in balsam or HSR.

D. PROPYLENE GLYCOL-SUDAN BLACK TECHNIQUE

This procedure is recommended for the demonstration of mitochondria, phospholipids as well as neutral fats in monolayers. Lipids are stained blue-black.

MATERIALS

1. Sudan Black Solution

Sudan Black B	0.7 gm	Dissolve Sudan Black in propylene glycol at 100-110°C. Do not exceed this temperature. Filter hot through Whatmann No. 2 filter paper. Cool and refilter with vacuum through medium porosity fritted glass filter.
Absolute propylene glycol	100.0 ml	

* Allied Chemical & Dye Corp. (C.I. 42053)

2. 85% propylene glycol

3. Polyvinylpyrrolidone (PVP) Mounting Medium (38)

(Refractive index 1.43)

Polyvinylpyrrolidone*	50 gm	Dissolve PVP in distilled water, let stand overnight. Add glycerol and stir mixture thoroughly. A crystal of thymol can be added as a preservative.
Distilled water	50 ml	
Glycerol	2 gm	
Thymol, crystalline		

PROCEDURE

1. Fix in cold neutral buffered formalin (p. 177) for 10 minutes.
2. Rinse in water for 2-5 minutes.
3. Rinse in propylene glycol for 3-5 minutes.
4. Stain in Sudan Black B for 5-7 minutes.
5. Destain in 85% propylene glycol for about 2-3 minutes.
6. Wash in distilled water for 3-5 minutes.
7. Mount with polyvinylpyrrolidone.

NOTE: Cold acetone extraction for 30 minutes prior to this procedure will remove all lipids other than phospholipid.

E. BAKER ACID HEMATIN FOR PHOSPHOLIPID

MATERIALS

1. Formol-Calcium fixative

Formalin (40% Formaldehyde)	10 ml	Mix ingredients in distilled water. Add excess of $MgCO_3$ to neutralize.
10% aqueous $CaCl_2$ (anhydrous)	10 ml	
Distilled water	80 ml	

*Type NP-K 30 Polyvinylpyrrolidone. Antara chemicals, Division of General Aniline and Film Corp., New York, New York.

2. Dichromate-Calcium solution

Potassium dichromate	5 gm	Dissolve ingredients in distilled water. Solution is stable. A small precipitate may be ignored.
Calcium chloride (anhydrous)	1 gm	
Distilled water	100 ml	

3. Baker acid hematin

Hematoxylin	50 mg	Dissolve the hematoxylin in distilled water. Add the iodate and heat just to a boil. Cool and then add glacial acetic. Prepare on same day of use.
Potassium or Sodium iodate (1% solution)	1 ml	
Glacial acetic acid	1 ml	
Distilled water	48 ml	

4. Borax-ferricyanide solution

Potassium ferricyanide	0.25 gm
Borax ($Na_2B_4O_7 \cdot 10\ H_2O$)	0.25 gm
Distilled water	100 ml

5. Weak Bouin fixative

Saturated aqueous picric acid	50 ml	Add the glacial acetic acid just before use.
Formalin (40% formaldehyde)	10 ml	
Distilled water	35 ml	
Glacial acetic acid	5 ml	

6. Pyridine, USP

PROCEDURE

1. Rinse cover slips in BSS (p. 214).
2. Fix in formol-calcium for 18 hours.
3. Incubate in dichromate calcium solution for 18 hours at 28° C.
4. Incubate in dichromate calcium solution for 24 hours at 60° C.
5. Wash in distilled water for 5 minutes, changing water repeatedly.
6. Stain in freshly prepared Bakers acid hematin for five hours at 37° C.
7. Rinse in distilled water.
8. Differentiate in a Borax-ferricyanide solution for 18 hours at 37° C.
9. Rinse repeatedly in distilled water.
10. Mount in Polyvinylpyrrolidone (p. 192).

BAKER ACID HEMATIN CONTROL

1. Rinse coverslips in BSS.
2. Fix in weak Bouin for 20 hours.
3. Wash coverslips in water (3 changes, 5 minutes each).
4. Place in pyridine solution for 1 hour at 28° C.
5. Place in fresh pyridine solution for 24 hours at 60° C.
6. Wash cover slips in dichromate calcium solution for 18 hours at 28° C.
7. Proceed as in Baker Acid Hematin Test for phospholipids (Step #5).

F. ALKALINE PHOSPHATASE TECHNIQUE

This histochemical procedure is included as one example of many such techniques that are becoming increasingly available. The procedure recommended here is the azo-dye method in view of its simplicity (39). Sites of enzyme activity appear as dark blue granules.

MATERIALS

1. Reaction mixture

Naphthyl ASE phosphate*	20 mg	Dissolve the naphthyl ASE phosphate in dimethyl formamide. Add 50 ml of the buffer. Add $MgSO_4$ Add rest of buffer. Then Add Blue RR and shake thoroughly.
N, N-dimethyl formamide	0. 5 ml	
Fast Blue RR**	60 mg	
$MgSO_4$ anhydrous	60 mg	
0. 1M Tris Buffer***(pH 8. 8)	100 ml	

2. PVP mounting medium (p. 192).

PROCEDURE

1. Fix for 5 minutes in cold neutral buffered formalin (p. 176).
2. Rinse in distilled water 1-2 minutes.
3. Incubate in freshly prepared reaction mixture for 20 minutes.
4. Rinse in distilled water.
5. Mount coverslip on slide with PVP.

NOTE: As a control, use the diazonium salt for 20 minutes without substrate.

* Nutritional Biochemicals Inc., Cleveland, Ohio
** 4 Benzoylamino-3, 5-Dimethoxyaniline, C.I., 37155, Dajac Laboratories, Philadelphia, Pa.
*** Sigma 7-9; Sigma Chemical Co., St. Louis, Mo.

G. NON-SPECIFIC ESTERASE TECHNIQUE

This is another histochemical procedure for a group of enzymes that is being increasingly studied. Sites of enzymatic localization stain red.

MATERIALS

1. Reaction mixture

Absolute propylene glycol	20 ml	Dissolve dye (Garnet G. B. C.) in distilled water. Dissolve substrate in buffer. Mix and add remaining ingredients.
Sorensen phosphate buffer pH 7.6	20 ml	
12 ml of M/15 $NaH_2PO_4 \cdot H_2O$ and		
88 ml of M/15 Na_2HPO_4 (anhydrous)		
Distilled water	58 ml	
1% naphthyl AS-Acetate in 6% acetone (substrate)	2 ml	
Garnet G. B. C. salt*	50 mg	

2. Polyvinylpyrrolidone (PVP) (p. 192).

NOTE: As a control incubate without substrate.

PROCEDURE

1. Fix in cold neutral buffered formalin for 5 minutes.
2. Rinse in distilled water for 1-2 minutes
3. Incubate in substrate for 15 minutes.
4. Rinse in distilled water.
5. Mount in PVP.

H. LACTIC DEHYDROGENASE

The demonstration of lactic dehydrogenase (LDH) activity in cultivated cells is provided as a representative example of the general procedure for demonstrating pyridine nucleotide linked oxidative enzymes (40).

MATERIALS

1. Monolayer of cells on coverslip (9 x 22 mm).
2. Ten ml Hanks BSS (p. 214).
3. Two columbia staining jars.

* General Dyestuff Co., General Aniline and Film Corp., N.Y.

4. Lactate stock solution

Sodium (DL) lactate	5.0 gm	Dissolve in 5 ml distilled water. Adjust pH to 7.4 with N NaOH. Bring to final volume of 10 ml. Store in refrigerator.

5. Cyanide solution

Sodium cyanide	4.0 gm	Dissolve in 5 ml distilled water. Adjust pH with N HCl to pH 7.2. MUST BE DONE IN VENTILATED HOOD! HIGHLY POISONOUS GAS EVOLVES. Bring to 10 ml final volume. Store in refrigerator.

6. PMS solution

Phenazine methosulfate*	80 mg	Dissolve in 100 ml distilled water. Store in dark bottle in refrigerator.

7. Reaction mixture

Diphosphopyridine nucleotide (DPN)**	2.5 mg	Just before use, dissolve DPN and Nitro BT in Phosphate buffer. Add 0.1 ml of lactate stock solution and 0.1 ml of cyanide solution; <u>immediately</u> before use add 0.1 ml of PMS solution
Nitro B. T.***	4.0 mg	
Sorenson's phosphate Buffer (0.1M, pH 7.4)	8 ml	

8. Ten ml neutral buffered formalin (p. 177).

9. PVP Mounting medium (p. 192).

* Sigma Chemical Co., St. Louis, Mo.

** California Corporation for Biochemical Research, Los Angeles, Calif.

*** 2,2' di-p-nitrophenyl-5,5' diphenyl-3,3'-(3,3'-dimethoxy- 4,4' biphenylene) ditetrazolium chloride. Sigma Chemical Co., St. Louis, Mo.

PROCEDURE

1. Remove coverslips from culture tubes, rinse in warm BSS.
2. Incubate in freshly prepared Reaction mixture for 5-15 minutes.
3. Rinse briefly in distilled water.
4. Fix in 10% neutral buffered formalin for 10 minutes.
5. Rinse in distilled water.
6. Mount on slides with PVP.

ENUMERATION OF CHROMOSOMES

Analysis of the chromosomal complement of cells in culture is becoming an increasingly important tool in genetic studies as well as for the characterization of cell lines (41-49). Characterization on the basis of chromosome number alone is not reliable generally since most cell lines become heteroploid in culture. In several instances morphologically distinctive chromosomes have emerged in a cell population which serve to characterize the cell line. The distribution of metacentrics, telocentrics, minutes, *etc.* also serves to characterize cell populations.

A certain amount of experience with the procedures of chromosomal analysis is required before reliable results can be obtained. It is wise to check your technique with material which has been characterized adequately.

MATERIALS

1. Monolayer culture of L-M strain mouse cells in 199 peptone (p. 11). These cells should be in early or mid log phase of growth.
2. Sterile colchicine* (0.1 mg/ml in BSS). Sterilize by filtration and store at 4°C.
3. Sterile hypotonic solution (1% sodium citrate in distilled water). Prepare fresh and use warm at 37° C.
4. One 15 ml conical centrifuge tube.
5. One sterile 1 ml cotton plugged serological pipette.
6. Two 10 ml and four 5 ml serological pipettes.
7. One rubber policeman.
8. Acetic acid-methanol fixative (1 part glacial acetic acid and 3 parts absolute methanol).
9. Seventy per cent ethanol.
10. Acetic Orcein stain (Dissolve 0.5 grams of Gurr natural orcein** in 100 ml of boiling 45% acetic acid, cool and filter 3 times). Filter before use if crystals have formed.

* Colchicine Alkaloid USP, Fischer Scientific Co.

** Gurr's Limited, London, England.

11. Microscope slides (1x3") and 22 mm square coverslips.

12. Bibulous paper.

13. Small metal spatula.

14. Krönig cement*.

PROCEDURE

1. With a sterile 1 ml serological pipette add 0.1 ml of colchicine solution to a monolayer culture of L-M cells in early log growth phase. Incubate at 35°C for eight hours!

2. After eight hours incubation, harvest the monolayer by scraping with a rubber policeman (p. 12).

3. With a 10 ml serological pipette triturate 3-4 times to disperse the cells and transfer the cell suspension to a 15 ml conical centrifuge tube.

4. Centrifuge the cells gently (150G) for 5 minutes and pipette off the medium.

5. Add 3 ml of warm sodium citrate solution and resuspend. Let stand for 6-8 minutes at 37° C, then add 1 drop of acetic-methanol fixative.

6. Centrifuge for 5 minutes at 150G and pipette off the supernatant fluid. Resuspend the cells in 1 ml of acetic acid-methanol fixative for each ml of the original suspension and let stand at room temperature for 10 minutes.

7. Centrifuge and change the fixative two more times. Finally resuspend in fixative.

8. Dip a clean glass slide in 70% ethanol and immediately drop one or two drops of cells suspended in fixative on the wet slide and pass it quickly through a flame to ignite. (This will spread the chromosomes). Blow to complete drying. Allow the slide to stand for 10-15 minutes before proceeding to the next step.

* Fisher Scientific Co.

9. Add one or two drops of acetic-orcein stain to the slide and gently lower a coverslip onto the stain taking care not to introduce bubbles. Gently blot between bibulous paper and seal the edges with Krönig cement, applied with a heated metal spatula.

10. Examine the stained preparation with an oil immersion phase contrast objective. Select a chromosome spread which is well separated from other cells but with the cell membrane intact to insure that you are dealing with a single cell. Count the number of chromosomes. Repeat for at least 10 additional cells. (For reliable statistics it is necessary to examine 50-100 spreads).

11. With a camera lucida trace the outlines of several of the chromosomes in a well spread preparation.

PHASE AND INTERFERENCE MICROSCOPY

Observation of specimens with the light microscope is dependent upon the relative absorption of light, a function of the thickness of the specimen and the refractive indices of the specimen and mounting medium (optical path). As a consequence, cells or tissues examined with the light microscope are best visualized when they have been fixed and stained to emphasize specimen contrast. Bright field, dark field or polarized light (for doubly refractive materials) have been employed for the study of living cells where specimen details consist of small differences in optical path. Although these methods provide contrast between the preparation and the mounting medium, they fail to produce contrast within the specimen. Vital staining techniques also may be used but this procedure is time consuming and may cause severe distortion. The phase-contrast microscope and the interference microscope (50-52) permit visualization of living cells.

The phase-contrast instrument is equipped with phase retardation plates which emphasize the refractive differences between microscopic objects and the medium in which they are immersed. Light rays traverse equal distances through air, water, and glass at different rates and emerge out of phase. These phase differences are not detectable by the eye which responds primarily to intensity differences. The phase-contrast microscope converts these phase differences into intensity differences thus rendering the various indices of refraction visible. Phase differences are enhanced by the choice of a mounting medium with a low index of refraction, such as saline solutions, and the instrument is therefore particularly adapted for studying living tissues.

Interference microscopy can be accomplished by the recombination in the image of two beams of light from the same source, one having been modified by passing through the specimen and the other through the surrounding medium. This microscope provides variable color contrast with white light illumination and intensity variations with monochromatic light. Thus, interference microscopy can be used to measure the thickness of the optical path and mass of the specimen. For purely observational purposes, the phase-contrast microscope is recommended whereas interference microscope is essential for accurate measurement of phase changes which permits measurement of mass, optical path, *etc.*

Cells from monolayers or suspension cultures may be examined satisfactorily in normal balanced salt solution with the phase or the interference microscope. The thickness of specimen which can be examined depends on its transparency and its optical density. Dense tissues or materials up to about 4 microns thick may be used and less dense specimens with a thickness of 100 microns or more can be studied. Large specimens such as explants must be embedded and sectioned with a microtome (53, 54). Coverglasses about 0. 18 of a mm thick are essential for best results as objectives in most phase and interference microscopes are corrected to this standard. A box of #1$\frac{1}{2}$ coverslips will contain more 0.18 mm coverglasses than will a box of #1 or #2 coverslips. It should be recognized also that hollow ground slides (unless ground with a flat surface) and hanging drop mounts will act as lens systems and therefore are not suitable for phase or interference microscopy. Special quartz slides may be used. Similarly, the split-beam interference microscope reveals the total optical path which includes the slide, the mounting medium, the specimen and the coverglass. Any unevenness, defects in the glass, or inhomogeneity in the mounting medium will produce artefacts in the preparation. Some distortion (wedge effect) is usually present in the preparation even with better than average quality coverglasses and slides and when care is used in mounting the specimen. The distortion in interference microscopy will manifest itself as a change of luminescence as the stage is turned or in a different reading depending on the orientation of the sample.

PREPARATION OF SERUM AND PLASMA *

Several generalizations concerning the use of serum for growth of cells can be made. Homologous serum provides best growth of cells or organs in culture. If heterologous serum must be used, cytotoxicity increases in proportion to the concentration of serum employed. Freshly drawn serum is superior to serum stored for weeks at 4 or -20° C. It is most economical to use commercial serum for routine cell culture techniques. Lyopholized sera, because of the loss of CO_2, is alkaline and often requires gassing with CO_2 to bring it to a physiological pH, *viz.*, 7.2 or 7.4. Because of cytotoxicity or nonspecific virus neutralization, contributed by the "normal" antibody-complement system, it may be prudent to heat-inactivate sera (56° C for 30 minutes) prior to use. Some of the variations in the growth qualities of sera are known to be attributable to differences among individual donors or/and to seasonal or dietary factors which are difficult to control. Because of these contingencies sera may be pooled to minimize such variations. Recent knowledge concerning the selective action of sera on variants in cell populations (55) suggests caution in choice of serum where population changes may be of importance.

Plasma clot procedures still are used in a number of laboratories for establishing explant cultures or for deriving established cell strains. Since many commercial preparations of heparin contain phenol or some other toxic preservative, heparin from Connaught Medical Research Laboratories (Toronto, Canada) is recommended.

MATERIALS (human venipuncture)

NOTE: State laws may require supervision of bleeding of human subjects by an attendant physician.

1. One sterile 30 ml hypodermic syringe.
2. Two sterile 19 gauge, 2 inch, short bevel hypodermic needles.
3. Cotton plegets moistened with 70% ethanol for skin disinfection.
4. Cotton plegets moistened with tincture of iodine (2%) for skin disinfection.
5. One flexible rubber hose to be used as a tourniquet.
6. One sterile, screw-cap, 40 ml, conical centrifuge tube.
7. One sterile gauze pad to be used as a compress.
8. Media for sterility checks (p. 7).
9. One "propipette"**.
10. Three sterile 16 x 125 mm screw-cap tubes.
11. Sterile applicator sticks.

* Because cell cultures can be contaminated with foreign cells, careful preparation of serum or plasma to insure that they are cell free is required.

** Instrumentation Associates, New York City.

MATERIALS (to obtain cockeral blood)

1. One sterile, 30 ml syringe (oiled before autoclaving with USP mineral oil, 4% in petroleum ether; alternatively, syringes may be rinsed with heparin (0.2 mg%) prior to use to prevent clotting).
2. Two sterile 20 gauge, 3 inch, short bevel hypodermic needles.
3. Plegets as described above.
4. One sterile, screw-cap, 40 ml conical centrifuge tube containing 0.3 mg of heparin.
5. One sterile gauze pad to be used as a compress.
6. One "propipette"*.
7. Media for sterility checks (p. 7).
8. Sterile applicator sticks.
9. Three, sterile, 16 x 125 mm screw-cap tubes.

PROCEDURE

1. Carefully disinfect the skin and draw 30 ml of human blood by venipuncture (median cubital vein) as demonstrated by the instructor. Aseptically transfer the blood to the centrifuge tube, permit it to clot, and free the clot from the walls of the tube with a sterile applicator stick.

2. Bleed the rooster from the wing vein or by cardiac puncture as demonstrated. (The head of the bird should be covered to minimize restlessness; a sharp needle and a good light source are needed to facilitate entry into the vein; blind probing is worse than useless.) Mix the blood and heparin thoroughly to prevent clotting. Press the gauze pad over the site of puncture and <u>hold it firmly in place</u> for at least five minutes; avian blood clots slowly.

3. Centrifuge all specimens for 30 minutes at approximately 2500 RPM; aspirate supernatant fluids aseptically by use of a 10 ml pipette and a rubber bulb. A "propipette" bulb permits ready removal of supernatant fluids without resuspending cells.

4. Dispense the serum and plasma in 10 ml aliquots into the 16 x 125 mm tubes.

5. Inoculate culture media for sterility checks (p. 7).

6. Store the serum and plasma at 4° C.

* Instrumentation Associates, New York City.

PREPARATION OF EMBRYO EXTRACT

Although established cell strains propagate readily in balanced salt solutions to which serum has been added, it may be expedient if not essential to supplement culture media with growth stimulatory substances for isolation of cell strains or to establish explant or organ cultures. Extracts of tissues from a wide range of animal sources have been used for this purpose. Because embryonic tissue usually is rich in growth stimulatory substances, it is used in preference to extracts of adult tissues. If tissue differentiation is germane to experimental studies, age of the embryos used for preparation of extracts must be considered, *e.g.*, extracts from six to nine day old chicken embryos significantly promote growth while those from twelve to fourteen day old embryos influence tissue differentiation and growth. Chicken embryos from nine to fourteen days of age are the most convenient source material for preparation of embryo extract. When large quantities of material are required, or if mammalian embryos are employed, tissues can be minced in a Waring type blendor. Because embryo extracts can serve as a source of somatic cells that may contaminate cell strains, it may be necessary to filter such additives as a precautionary measure (56).

Embryo extract may serve as a source of bacterial contaminants, *viz.*, Salmonellae, or may contain viruses or other filterable microbes, *e.g.*, PPLO (57) which are not readily detectable. Effective use of embryo extracts in tissue culture research requires an appreciation of these problems as they relate to interpretation of experimental findings.

MATERIALS

1. Three, nine to twelve day old embryonated chicken eggs.
2. Waxed egg carton to hold eggs upright.
3. Cotton plegets moistened with 2% tincture of iodine.
4. Cotton plegets moistened with 70% ethanol.
5. Two fine-tip forceps.
6. Two sterile Petri dishes.
7. One sterile, 30 ml syringe, the body equipped with a 28 gauge stainless steel wire mesh; the body and plunger of the syringe should be wrapped separately and autoclaved.
8. One sterile screw-cap, 40 ml conical centrifuge tube.
9. Three, sterile, screw-cap, 16 x 125 mm tubes.
10. Egg candler.
11. Fifty ml BSS (p. 214).
12. Culture media for sterility checks (p. 7).

PROCEDURE

1. Candle the eggs and select those which are fertile.

2. Place the fertile eggs, air-sac end uppermost, in the waxed egg cartons.

3. Disinfect the egg shells with the iodine and ethanol plegets as demonstrated by the instructor. Gently crack the shell over the air-sac; remove the shell surrounding the air-sac with sterile forceps.

4. Tear off the shell membrane, gently grasp the neck of the embryo and exert gentle upward traction until the embryo is free of attached membranes. (If forceps are compressed too tightly or if excessive traction is applied, the neck will break and the embryo will be lost.)

5. Transfer the embryos to a Petri dish or directly to the 30 ml syringe.

6. Insert the plunger into the syringe, measure the compressed volume of the pooled embryos, and then forcibly extrude the embryos through the screen.

7. Collect the minced embryonic tissue in the conical centrifuge tube and aseptically add an equivalent volume of BSS.

8. Mix the minced tissue and BSS thoroughly; centrifuge the embryo extract for 30 minutes at approximately 2500 RPM.

9. Dispense the supernatant fluid into the 16 x 125 mm tubes and inoculate culture media for sterility checks (p. 7). Store the embryo extract at -20° C.

NOTE:

1. Thawed embryo extract usually must be centrifuged before use and the sedimentable material discarded.

2. Fresh, frozen or lyophilized embryo extract can be used to coagulate plasma; ultrafiltrates of embryo extracts are not satisfactory for this purpose.

PRESERVATION OF CELLS

For many years the only method for maintaining tissue cell lines was by serial subculture at frequent intervals. By this method Carrel maintained his celebrated culture of chick heart fibroblasts for over 34 years (58). Several groups of workers (59-64) have demonstrated the feasibility of long term storage of animal cells at low temperatures. This has greatly simplified the problem of cell preservation. A variety of primary cells, cell strains and established cell lines have been shown to survive when stored at -65° C or below without a discernible change of properties.

While satisfactory survival has been attained with storage in the range of -65° C (electric deep freeze) to -79° C (solid carbon dioxide chest), it is recommended that the storage temperature not be above -90° C. Storage in liquid nitrogen (-196° C) or in the vapour above liquid nitrogen (-150° C to -180°C) are the most practicable methods currently available for achieving such temperatures.

There are relatively few hazards involved in working with liquid nitrogen. Direct contact with the skin is to be avoided as it may result in frostbite. Work with nitrogen should always be in a well ventilated room to avoid anoxia though the danger is not great under ordinary circumstances. While liquid nitrogen is not explosive *per se*, a danger is involved when containers are immersed in liquid nitrogen. Unless all immersed containers are watertight, liquid nitrogen may leak in and when the containers are removed from storage, vaporization of the nitrogen will lead to rapid expansion with a resulting explosion of the container. This is most likely to occur with ampoules which are improperly sealed or with stoppered or screw-cap containers. Storage in the vapor phase prevents such an occurrence.

Adequate attention must be given to procedures for freezing and thawing and to the menstrum in which the cells are suspended. The critical points in technique are: 1) slow freezing, 2) rapid thawing, 3) use of 5-20% (v/v) glycerol or 5-10% (v/v) dimethyl sulfoxide in the freezing and storage medium, and 4) storage at temperatures below -90° C.

The freezing and storage medium normally is the growth medium for the particular cells being frozen. To this is added 5-20% glycerol or 5-10% dimethyl sulfoxide. The exact amount added will depend both upon the particular cells and upon the amount of serum in the medium. Freezing should take place at approximately 1°C drop per minute down to -30°C. The temperature can then be dropped rapidly by placing the sample in liquid N_2 or in the vapor above liquid N_2. Thawing should be done in a water bath at +45°C with agitation and should require less than 1 minute.

Glycerol or dimethyl sulfoxide should be added to the culture medium before suspending the cells in the medium or should be added gradually to the cell suspension. Upon thawing, the cell suspension should be diluted to the desired concentration of cells by addition of fresh medium. This should be done gradually. The final concentration of glycerol or dimethyl sulfoxide should be less than 1%. Under no circumstances should the cell suspension be centrifuged before plating as this will cause unnecessary trauma.

Several methods are available for attaining either an accurate freezing rate or a satisfactory approximation of a 1°C/minute drop in temperature. The most elaborate method involves the use of an electronic device for programming the temperature drop by controlling the rate at which liquid nitrogen vapor is introduced into a sample storage chamber.* A less sophisticated method freezes samples in the vapor contained in the neck of a liquid nitrogen storage tank.* By regulating the position of the samples in the neck of the tank, different freezing rates can be achieved (64). A less accurate method is to place ampoules at 4°C (ordinary refrigerator) for 2-4 hours, then into the -20°C freezing compartment for 4-24 hours and finally into the storage chest.

Ampoules for freezing should be of borosilicate glass with a low coefficient of expansion.** Sealing should be done by the pull method, rather than by fusion, using an oxygen flame. An automatic pull sealer is available commercially.*** Ampoules may be tested for proper sealing by immersing them in strong dye solution or in 95% ethanol. In the latter case a leak will be indicated by precipitation of protein in the cell suspension.

If gummed labels are used these should be autoclaved on the ampoules to prevent loss on storage. If masking tape is used this procedure is not necessary. Ampoules may be filled with a hypodermic syringe using a large bore needle of adequate length (*e.g.* 15 gauge, $3\frac{1}{2}$").

A typical protocol for preparation of L-M mouse cells for freezing is presented below. With suitable modifications of medium and harvest procedures the method is applicable to many other cells.

MATERIALS

1. Five 2 oz. French square monolayer cultures of L-M cells in 199 peptone (p. 12). These cells should be in the mid-logarithmic phase of growth.

* Linde Division, Union Carbide Corp., N.Y., N.Y.
** Wheaton Glass Co., Millville, N.J.
*** Kohlenberg-Globe Equipment Co., Sarasota, Fla.

2. Five sterile rubber policemen.

3. Sterile reagent grade dimethyl sulfoxide (sterilize by autoclaving).

4. Sterile 40 ml screw-cap conical centrifuge tube.

5. Twenty ml sterile 199 peptone (p. 238).

6. Ten sterile 1 ml ampoules*, **.

7. One sterile 10 ml hypodermic syringe and one sterile 15 gauge, $3\frac{1}{2}$ inch hypodermic needle.

8. Six 5 ml, two 10 ml and two 1 ml sterile cotton-plugged serological pipettes.

PROCEDURE

1. Using sterile rubber policemen scrape the monolayers into the medium in which they are growing. Triturate with a pipette to disperse the cells and transfer the cell suspension to the sterile 40 ml centrifuge tube. Pool the cells from the five bottles.

2. Centrifuge the cell suspension at 500 RPM for 10 minutes and then decant the medium.

3. Add 10% (v/v) of dimethyl sulfoxide (DMS) to 20 ml of 199 peptone. Mix thoroughly.

4. Add 5 ml of the 199 peptone (containing DMS) to the sedimented cells and resuspend them. Count the cells and adjust the concentration to $2X10^6$/ml by diluting with 199 peptone (containing DMS) as required. Do a viable count using the erythrosin B method (p. 157) and record the results.

5. Take up the cell suspension in the sterile 10 ml hypodermic syringe fitted with a 15 gauge, $3\frac{1}{2}$ inch hypodermic needle. Dispense 1 ml of cell suspension into each sterile ampoule.

6. Seal the ampoules by pulling out the neck in an oxygen flame.

7. Label each ampoule with the proper strain designation, number of cells/ml, medium and preservative and the date.

Upon recovery of the cells from the freezer, total and viable cell counts should be made and compared with the counts at the time of freezing to determine percentage survival. With cell lines which clone well, more meaningful data can be obtained by determining plating efficiency (p. 52) before and after storage.

* Wheaton Glass Co., Millville, N.J.

** A small glass vial of appropriate size may be inverted over the neck of each ampoule during sterilization to protect against contamination during the process of filling and sealing.

TRANSPORT OF CELLS

Cells can be transported readily, even for great distances, if they are prepared properly. The means of shipping (air express, air mail, first class mail) will depend upon the distance of shipment and accessibility of transportation. The chief considerations are: (a) avoiding undue delay, (b) avoiding extreme changes of temperature, (c) relative cost. Cells can be refrigerated during shipment although generally this is not necessary unless long delays are anticipated. Tubes or bottles should be packed in insulated containers both to protect against breakage and to prevent extreme changes in temperature. Cells being made ready for shipment should be fed one day prior to shipping and should be actively proliferating.

Monolayer cultures may be shipped by filling the culture vessel (usually a tube) completely with medium. This avoids splashing which may wash cells off the glass. Tight fitting caps or stoppers are necessary. An alternative method is to remove the medium, which is placed in a separate container and shipped with the cells. Upon arrival the medium is replaced and the cells are incubated until a complete monolayer is formed.

Suspensions of cells may be shipped quite easily. If cells adhere to the vessel during transport they may be scraped off or trypsinized as indicated. Upon arrival the cells are planted according to routine procedure. Suspensions of cells in the frozen state may also be shipped when packed in dry ice. Insulated containers, which are readily available, make this feasible even for long distances.

REFERENCES

1. HARRIS, M., Growth measurements on monolayer cultures with an electronic cell counter, Cancer Research, 19: 1020-1024 (1959).
2. HANKS, J. H., and WALLACE, J. H., Determination of cell viability, Proc. Soc. Exper. Biol. & Med., 98: 188-192 (1958).
3. PHILLIPS, H. J., and TERRYBERRY, J. E., Counting actively metabolizing tissue cultured cells, Exp. Cell Res., 13: 341-347 (1957).
4. SANFORD, K. K., EARLE, W. R., et al., The measurement of proliferation in tissue cultures by enumeration of cell nuclei, J. Nat. Cancer Inst., 11: 773-795 (1951).
5. WAYMOUTH, C., A rapid quantitative hematocrit method for measuring increase in cell population of strain L (Earle) cells cultivated in serum-free nutrient solutions, J. Nat. Cancer Inst., 17: 305-313 (1956).
6. MUNYON, W. H., and MERCHANT, D. J., The relation between glucose utilization, lactic acid production and utilization and the growth cycle of L strain fibroblasts, Exper. Cell Res., 17: 490-498 (1959).
7. KUCHLER, R. J., and MERCHANT, D. J., Growth of tissue cells in suspension, Univ. Michigan M. Bull., 24: 200-212 (1958).
8. OYAMA, V. I., and EAGLE, H., Measurement of cell growth in tissue culture with a phenol reagent (Folin-Ciocalteau), Proc. Soc. Exper. Biol. & Med., 91: 305-307 (1956).
9. PAUL, J., The chemical determination of deoxyribonucleic acid in tissue cultures, J. Biophys. & Biochem. Cytol., 2: 797-798 (1956).
10. McINTIRE, F. C., and SPROULL, M. F., A simple method for determination of desoxypentose nucleic acid in tissue cultures, Proc. Soc. Exper. Biol. & Med., 95: 458-462 (1957).
11. PAUL, J., Determination of the major constituents of small amounts of tissue, Analyst, 83: 37-42 (1958).
12. HUTCHISON, W. C., and MUNRO, H. N., The determination of nucleic acids in biological materials, a review, Analyst, 86: 768-813 (1961).
13. SCHMIDT, G., and THANNHAUSER, S. J., A method for the determination of desoxyribonucleic acid, ribonucleic acid, and phosphoproteins in animal tissues, J. Biol. Chem., 161: 83-89 (1945).
14. KABAT, E. A., Kabat and Mayer's Experimental Immunochemistry, 2nd ed., Springfield, Ill., Charles C. Thomas, (1961).
15. KOCH, F. C., and McMEEKIN, T. L., A new direct nesslerization micro-Kjeldahl method and a modification of the Nessler-Folin reagent for ammonia, J. Am. Chem. Soc., 46: 2066-2069 (1924).
16. CAMPBELL, D. H., GARVEY, J. S., et al., Methods in Immunology, New York, W. A. Benjamin, Inc., (1963).
17. LOWRY, O. H., ROSEBROUGH, N. J., et al., Protein measurement with the folin phenol reagent, J. Biol. Chem., 193: 265-275 (1951).
18. CROXTON, F. E., Elementary statistics with applications in medicine and the biological sciences, New York, Dover Publications, Chapter 6 (1953).
19. MEJBAUM, W., Uber die Bestimmung kleiner Pentosemengen, insbesondere in Derivaten der Adenylsaure, Hoppe-Seyler's Ztschr, f. physiol. Chem., 258: 117-120 (1939).
20. BURTON, K., A study of the conditions and mechanisms of the diphenylamine reaction for the colorimetric estimation of deoxyribonucleic acid, Biochem. J., 62: 315-323 (1956).
21. KOPRIWA, B. M., and LEBLOND, C. P., Improvements in the coating technique of radioautography, J. Histochem., 10: 269-284 (1962).
22. PRESCOTT, D. M., (ed.), Methods In Cell Physiology, New York, Academic Press, Vol. I., (1964).
23. STANNERS, C. P. and TILL, J. E., DNA Synthesis in individual L-strain mouse cells, Biochem. et biophys. acta, 37: 406-419 (1960).
24. WHITMORE, G. F., STANNERS, C. P., et al., Nucleic acid synthesis and the division cycle in x-irradiated L-strain mouse cells, Biochem. et biophys. acta, 47: 66-77 (1961).
25. PRESCOTT, D. M., Symposium: Synthetic processes in the cell nucleus. II. Nucleic acid and protein metabolism in the macro-nuclei of two ciliated protozoa., J. Histochem., 10: 145-153 (1962).
26. HSU, T. C., DEWEY, W. C., et al., Radiosensitivity of cells of Chinese hamster in vitro in relation to the cell cycle, Exp. Cell Res., 27: 441-452 (1962).

27. BAKER, J. R., Principles of Biological Microtechnique, New York, John Wiley & Sons, Inc., (1958).
28. GALLIGHER, A. E. and KOZLOFF, E. S., The Essentials of Practical Microtechnique, Philadelphia, Lea & Febiger, (1964).
29. McCLUNG, C. E., (ed.), Handbook of Microscopical Technique, 2nd ed., New York, Paul B. Hoeber, Inc. (1937).
30. ENDERS, J. F., and PEEBLES, T. C., Propagation in tissue cultures of cytopathogenic agents from patients with measles, Proc. Soc. Exper. Biol. & Med., 86: 277-286 (1954).
31. REISSIG, M., HOWES, D. W., et al., Sequence of morphological changes in epithelial cell cultures infected with poliovirus, J. Exper. Med., 104: 289-304 (1956).
32. GOMORI, G., Microscopic Histochemistry, Chicago, University of Chicago Press (1952).
33. LILLIE, R. D., Histopathologic Technic and Practical Histochemistry, New York, The Blakiston Co., Inc., (1954).
34. PEARSE, A. G. E., Histochemistry, Little, Brown and Co., Boston (1960).
35. COWDRY, E. V., Microscopic Technique in Biology and Medicine, Baltimore, Williams & Wilkins Co. (1943).
36. HOTCHKISS, R. D., A microchemical reaction resulting in the staining of polysaccharide structures in fixed tissue preparations, Arch. Biochem., 16: 131-141 (1948).
37. MOWRY, R. W., The special value of methods that color both acidic and vicinal hydroxyl groups in the histochemical study of mucins. With revised directions for the colloidal iron stain, the use of alcian blue G8X and their combinations with the periodic acid-schiff reaction. Ann. New York Acad. Sc., 106: 402-423 (1963).
38. BURSTONE, M. S., Polyvinylpyrrolidone as a mounting medium for stains for fat and for azo-dye procedures, Am. J. Clin. Path., 28: 429-430 (1957).
39. BURSTONE, M. S., Enzyme Histochemistry, New York, Academic Press, (1962).
40. CONKLIN, J. L., DEWEY, M. M., et al., Cytochemical localization of certain oxidative enzymes, Am. J. Anat., 110: 19-27 (1962).
41. LEVAN, A., Chromosome studies on some human tumors and tissues of normal origin, grown in vivo and in vitro at the Sloan-Kettering Institute, Cancer, 9: 648-663 (1956).
42. BERMAN, L., STULBERG, C. S., et al., Human cell culture. Morphology of the Detroit strains, Cancer Res., 17: 668-676 (1957).
43. HSU, T. C., and MOORHEAD, P. S., Mammalian chromosomes in vitro. VII. Heteroploidy in human cell strains, J. Nat. Cancer Inst., 18:463-471 (1957).
44. CHU, E. H. Y., and GILES, N. H., Chromosomal studies of normal and mutant clonal derivatives of human cell strains growing in vitro. (Abstract) Genetics, 42: 365-366 (1957).
45. FORD, D. K., and YERGANIAN, G., Observations on the chromosomes of Chinese hamster cells in tissue culture, J. Nat. Cancer Inst., 21: 393-425 (1958).
46. HSU, T. C., and KLATT, O., Mammalian chromosomes in vitro. IX. On genetic polymorphism in cell populations, J. Nat. Cancer Inst., 21: 437-473 (1958).
47. CHU, E. H. Y., SANFORD, K. K., et al., Comparative chromosomal studies on mammalian cells in culture. II. Mouse sarcoma-producing cell strains and their derivatives, J. Nat. Cancer Inst., 21: 729-751 (1958).
48. HSU, T. C., and MERCHANT, D. J., Mammalian chromosomes in vitro. XIV. Genotypic replacement in cell populations, J. Nat. Cancer Inst., 26: 1075-1083 (1961).
49. MOORHEAD, P. S., Chromosome morphology as a genetic marker, in Merchant, D. J. and Neel, J. V., (eds.), Approaches to the Genetic Analysis of Mammalian Cells, University of Michigan Press, p. 28-46, (1962).
50. BENNETT, A. H., OSTERBERG, H., et al., Phase Microscopy; New York, John Wiley & Son (1951).
51. BARER, R., Phase-contrast, interference contrast, and polarizing microscopy. in Mellors, R. C., (ed.) Analytical Cytology, New York, The Blakiston Division of McGraw-Hill Book Co., Inc., Chap. 3 (1955).
52. BARER, R., Phase contrast and interference microscopy in cytology, in Oster, G. and Pollister, A. W., (eds.), Physical Techniques in Biological Research, New York, Academic Press, Vol. 3, Chap. 2 (1956).

53. RICHARDS, O. W., The Effective Use and Care of the Microscope, Buffalo, New York, American Optical Co. (1949).

54. RICHARDS, O. W., A-O Baker Interference Microscope, Buffalo, New York, American Optical Co. (1957)

55. MURPHY, W. H., BULLIS, C., et al., Effects of heterologous sera on the modal distribution of variants in four strains of human epithelial cells, Cancer Res., 22: 906-913 (1962).

56. WAYMOUTH, C., and WHITE, P. R., Filtration of embryo extract for tissue cultures, Science, 119: 321-322 (1954).

57. ROTHBLAT, G. H., and MORTON, H. E., Detection and possible source of contaminating pleuro-pneumonialike organisms (PPLO) in cultures of tissue cells, Proc. Soc., Exper. Biol & Med., 100: 87-90 (1959).

58. CARREL, A., Present condition of a strain of connective tissue twenty-eight months old, J. Exper. Med., 20: 1-2 (1914).

59. SCHERER, W. F., and HOOGASIAN, A. C., Preservation at sub-zero temperatures of mouse fibroblasts (Strain L) and human epithelial cells (Strain HeLa), Proc. Soc. Exper. Biol. & Med., 87: 480-487 (1954).

60. STULBERG, C. S., SOULE, H. D., et al., Preservation of human epithelial-like and fibroblast-like cell strains at low temperatures, Proc. Soc. Exper. Biol. & Med., 98: 428-431 (1958).

61. SWIM, H. E., HAFF, R. F., et al., Some practical aspects of storing mammalian cells in the dry-ice chest, Cancer Res., 18: 711-717 (1958).

62. HAUSCHKA, T. S., MITCHELL, J. T., et al., A reliable frozen tissue bank: viability and stability of 82 neoplastic and normal cell types after prolonged storage at -78°C, Cancer Res., 19: 643-653 (1959).

63. SWIM, H. E., and PARKER, R. F., Preservation of cell cultures at 4°C, Proc. Soc. Exper. Biol. & Med., 89: 549-553 (1955).

64. KITE, J. H., Jr., and DOEBBLER, G. F., A simple procedure for freezing and storage of tissue cultures using liquid nitrogen, Nature, London, 196: 591-592 (1962).

Chapter VI.
MEDIA AND REAGENTS

BALANCED SALT SOLUTIONS

Each of the available balanced salt solutions (BSS) is basically a derivative of Ringer phosphate buffer as modified by Tyrode (1-6). Regardless of the formula, it should be recognized that a salt solution must be "balanced" in terms of quantity and ratio of ionic species so that it is physiologically acceptable in relation to pH, mineral content and osmotic pressure.

Five balanced salt solutions are presented in view of their widespread use. Hanks BSS is the most convenient in that the component solutions may be autoclaved separately and secondarily combined, negating the need for filtration. Dulbecco phosphate buffer has been used frequently for titration of viruses. Puck saline is employed commonly for cloning.

1. HANKS BALANCED SALT SOLUTION (BSS)

A. 10X SOLUTION

	Material	Amount	Preparation
Unit #1	$NaHCO_3$	3.5 gm	Dissolve in 250 ml distilled water. Dispense in a convenient bottle (50 ml screw-cap prescription bottle) and autoclave at 120° C for 15 minutes.
Unit #2	NaCl	80.0 gm	Dissolve in 800 ml distilled water.
	KCl	4.0 gm	
	$MgSO_4 \cdot 7H_2O$	2.0 gm	
	$Na_2HPO_4 \cdot 2H_2O$	0.6 gm	
	Glucose	10.0 gm	
	KH_2PO_4	0.6 gm	
Unit #3	$CaCl_2$	1.4 gm	Dissolve in 100 ml distilled water.

Unit #4	Phenol Red	0.4 gm	Mix phenol red in small amount of water until a paste, dilute to 150 ml with distilled water, titrate to pH 7 with N/20 NaOH. Make up to final volume of 200 ml. Preserve with 1-2 ml chloroform.

Add 100 ml of Unit #4 to Unit #2 and then add Unit #3 to make 1000 ml. Pour solution into glass stoppered bottle and add 3-4 ml chloroform as a preservative. This solution may be kept at room temperature for 6 months - 1 year.

NOTE: Minimize transfer of chloroform in preparation of the working solution. Be certain that bottle caps are loosened during autoclaving to insure that all chloroform is driven off.

B. WORKING SOLUTION (IX)

The working BSS is prepared by diluting 10X stock 1:10 with distilled water. Dispense in convenient size screw-cap bottles and autoclave at 120° C for 15 minutes. Aseptically add 2.5 ml of sterile sodium bicarbonate solution (Unit #1) to each 100 ml of BSS. The pH may be adjusted with CO_2. The balanced salt solution is now ready for use. Do not tighten caps until pH of BSS is 7.4.

2. EARLE BALANCED SALT SOLUTION (1X)

	Material	Amount	Preparation
Unit #1	NaCl	6.80 gm	Dissolve in 600 ml distilled water.
	KCl	0.40 gm	
	$MgSO_4$	0.10 gm	
	NaH_2PO_4	0.125 gm	
	$NaHCO_3$	2.20 gm	
	Glucose	1.00 gm	

NOTE: All amounts given are for the anhydrous salts. When hydrated forms are used, amounts must be adjusted. Dissolve, in order given, with agitation.

Unit #2	$CaCl_2$	10.0 gm	Dissolve in 100 ml distilled water.

For use, add 2.0 ml of Unit #2 to 200 ml distilled water and add to Unit #1. Make up to 1000 ml. Adjust to pH 7.4 with CO_2 and sterilize by filtration. Pressure filtration is preferred since less shift of pH results. Phenol red may be added as in Hanks BSS.

3. DULBECCO PHOSPHATE-BUFFERED SALINE (PBS-1X)

	Material	Amount	Preparation
Unit #1	NaCl	8.0 gm	Dissolve in 800 ml distilled water.
	KCl	0.2 gm	
	Na_2HPO_4	1.15 gm	
	KH_2PO_4	0.2 gm	
Unit #2	$CaCl_2$	0.1 gm	Dissolve in 100 ml distilled water.
Unit #3	$MgCl_2 \cdot 6H_2O$	0.1 gm	Dissolve in 100 ml distilled water.

Autoclave Units #1, 2, 3 separately and mix when cooled. Phenol red may be added as in Hanks BSS.

4. PUCK SALINE F (1X)

Material	Amount gm/liter	Preparation
NaCl	7.40	Dissolve in 1000 ml distilled water. Sterilize by filtration.
KCl	0.285	
$MgSO_4 \cdot 7H_2O$	0.154	
$CaCl_2 \cdot 2H_2O$	0.016	
$Na_2HPO_4 \cdot 7H_2O$	0.29	
KH_2PO_4	0.083	
$NaHCO_3$	1.20	
Glucose	1.10	
Phenol Red	0.0012	

5. PUCK SALINE G (1X)

(Recommended for use outside of CO_2 atmosphere)

Material	Amount gm/liter	Preparation
NaCl	8.0	Dissolve in 1000 ml distilled water. Sterilize by filtration.
KCl	0.4	
$MgSO_4 \cdot 7H_2O$	0.154	
$CaCl_2 \cdot 2H_2O$	0.016	
$Na_2HPO_4 \cdot 7H_2O$	0.29	
KH_2PO_4	0.150	
Glucose	1.10	
Phenol Red	0.0012	

TISSUE CULTURE MEDIA

1. CHEMICALLY DEFINED MEDIA

To date a defined medium has not been devised which will support the rapid proliferation of more than a limited number of cell lines. Nevertheless defined media have wide application as the basal constituent to which various undefined supplements are added. Seven representative media are presented.

A. MEDIUM 199 (Morgan, Morton and Parker, 7)

This medium was one of the earliest devised. Several revisions as well as the original formula continue in common use. It is complex in composition and presents serious problems in preparation. The formula listed below and the directions for preparation are from a widely used modification of the original method (8). One important change is the use of Hanks BSS rather than that of Earle as a base solution. This alteration permits the preparation of a 10X stock solution with less difficulty and also makes possible a wider range of pH control.

Modified Medium 199 (10X)

	Material	Amount	Preparation
Solution A	l-Arginine· HCl	700 mg	
	l-Histidine· HCl	200 mg	
	l-Lysine· HCl	700 mg	
	dl-Tryptophane	200 mg	
	dl-Phenylalanine	500 mg	
	dl-Methionine	300 mg	
	dl-Serine	500 mg	
	dl-Threonine	600 mg	
	dl-Leucine	1200 mg	
	dl-Isoleucine	400 mg	
	dl-Valine	500 mg	All of the ingredients of Solution A are added to the final 10X solution in dry form.
	dl-Glutamic Acid· H_2O	1500 mg	
	dl-Aspartic Acid	600 mg	
	dl-Alanine	500 mg	
	l-Proline	400 mg	
	Hydroxy-l-proline	100 mg	
	Glycine	500 mg	
	l-Glutamine	1000 mg	
	Sodium acetate· $3H_2O$	500 mg	

	Material	Amount	Preparation
Solution B	l-cystine l-tyrosine	200 mg 400 mg	Dissolve l-cystine and l-tyrosine in 200 ml of 0.075N HCl with gentle heating and vigorous shaking. Place in water bath at 37°C and shake at intervals until dissolved. Store at 25° C overnight. This solution is made fresh for each batch of 10X 199.
Solution C	Niacin Niacinamide Pyridoxine Pyridoxal Thiamin Riboflavin Ca-pantothenate i-inositol p-Aminobenzoic acid Choline chloride	25 mg 25 mg 25 mg 25 mg 10 mg 10 mg 10 mg 50 mg 50 mg 50 mg	Dissolve the dry ingredients in approximately 175 ml of distilled water and bring the volume to 200 ml with distilled water. Shake vigorously and stir with a stirring rod to ensure that all the ingredients dissolve. Sterilize by filtration and dispense in 2 ml amounts. Store at 5° C in the dark.
Solution E	d-Biotin	10 mg	Dissolve in approximately 75 ml of distilled water, add 1 ml of 1N HCl and bring to a volume of 100 ml with distilled water. Filter sterilize and dispense in 1 ml amounts. Store at 5° C.
Solution F	Crystalline folic acid	10 mg	Dissolve in 100 ml of 1X Hanks BSS. Filter sterilize and dispense in 1 ml amounts. Store at 5° C.

	Material	Amount	Preparation
Solution H-K	Crystalline Vitamin D_2 (Calciferol)	20 mg	Dissolve the vitamin D_2 in 40 ml of 95% ethanol. Add the cholesterol and stir vigorously to dissolve. Add 60 ml of 5% Tween 80. Filter sterilize and dispense in 5 ml amounts. Store at 25° C.
	Cholesterol	40 mg	
	Ethanol (95%)	40 ml	
	Tween 80 (5%)	60 ml	
Solution I	Vitamin E (dl alphatocopherol disodium phosphate)	10 mg	Dissolve in 100 ml of distilled water. (Forms a soapy solution) Filter sterilize and store in 1 ml amounts at 5° C.
Solution J	Vitamin K (Menadione)	10 mg	Dissolve in 100 ml of distilled water. Shake vigorously and/or incubate overnight at 37° C to dissolve. Filter sterilize, dispense in 1 ml amounts and store at 5° C.
Solution L	Adenine sulfate	100 mg	Dissolve in 12 ml of distilled water. Heat gently and add 1.25 ml of concentrated NH_4OH to speed the solution of the compound. Make fresh for each batch of 10X 199.
Solution M	Xanthine	100 mg	Dissolve the dry ingredients in 1,000 ml of distilled water containing 3.5 ml of concentrated NH_4OH. Heat gently and shake vigorously to dissolve. Sterilize by filtration and store in 30 ml amounts at 5 C.
	Hypoxanthine	100 mg	
	Thymine	100 mg	
	Uracil	100 mg	

	Material	Amount	Preparation
Solution M_1	Guanine hydrochloride	3 mg	Dissolve in 25 ml of distilled water, add 0.1 ml of concentrated NH_4OH and bring the volume to 30 ml with distilled water. Bring the solution to a boil and after cooling bring the volume back to 30 ml with distilled water. Prepare fresh for each batch of 10X 199.
Solution N	d-ribose d-deoxyribose	100 mg 10 mg	Dissolve in 10 ml of distilled water. Filter sterilize and store in 5 ml amounts at 5° C.
Solution O	Adenylic acid (muscle)	10 mg	Dissolve in 25 ml of distilled water. Filter sterilize and store at 5° C in 5 ml amounts.
Solution Q	$Fe(NO_3)_3 \cdot 9H_2O$	100 mg	Dissolve in 100 ml of distilled water. Add 1 drop of concentrated HNO_3. Filter sterilize and store at 5° C in 1 ml amounts.

Preparation of Solution D-G-P (labile compounds)

(Use 1.0 ml of solution D-G-P per liter of 1X 199)

	Material	Amount	Preparation
Solution D	l-cysteine HCl Glutathione Ascorbic acid	10 mg 5 mg 5 mg	Dissolve in 50 ml of distilled water.
Solution G	Crystalline Vitamin A Ethanol (95%) Tween 80 (5%)	10 mg 1 ml 10 ml	Dissolve Vitamin A in the ethanol and then add the Tween 80.

	Material	Amount	Preparation
Solution D-G			Combine Solutions D and G and bring the volume to 100 ml with distilled water. Filter sterilize and store at 5° C.
Solution D-G-P	Adenosine triphosphate (disodium salt 95% BTP)	200 mg	Dissolve in 20 ml of solution D-G. Filter sterilize and store at 5° C. This solution (D-G-P) must be used within 3 days.

Preparation of 1 liter of 10X 199

1. Dissolve 80 grams of NaCl in 200 ml of Solution B (prepared in advance).

2. Add Solutions M and M_1 (30 ml of each).

3. Add 4 grams of KCl and dissolve.

4. Dissolve 2 grams of $MgSO_4 \cdot 7H_2O$ in the solution. (Volume at this point with the washings from the beakers containing the various salts should be approximately 330 ml).

5. (a) Dissolve 0.6 grams of KH_2PO_4 in approximately 50 ml of distilled water. (b) Dissolve 0.6 grams of Na_2HPO_4 in the flask of 5 (a) and bring the volume to approximately 100 ml with distilled water.

6. Add the solution prepared in 5 (a and b) to the solution from step 4. Volume at this point should be approximately 425 ml.

7. Dissolve 10 grams of glucose in approximately 50 ml of distilled water. Add 0.2 grams of phenol red dye (soluble form) and bring the volume to approximately 100 ml with distilled water.

8. Add the solution prepared in 7 to the solution from step 6. (Volume at this point should be approximately 525 ml).

9. Dissolve 1.4 grams of anhydrous (12 mesh) $CaCl_2$ in approximately 175 ml of distilled water.

10. Add the $CaCl_2$ solution to the solution from step 8. Add the $CaCl_2$ slowly and with vigorous shaking and stirring of the solution in the flask. (This will prevent localized precipitation of insoluble $Ca_3(PO_4)_2$).

11. Add the following quantities of stock solutions C, E, F, H-K, I, J, N, O and Q to the solution from step 10 (total volume of additions = 22 ml).

Solution	Volume in ml
C	2
E	1
F	1
H-K	5
I	1
J	1
N	5
O	5
Q	1

Volume at this point, with the washings, should be approximately 775 ml.

12. Add the 12 ml of Solution L (freshly prepared) to the solution from step 11 and bring the volume in the flask to approximately 950 ml with distilled water.

13. Add to the flask of solution from step 12, in solid form, the amino acids and sodium acetate of "Solution" A. Shake and stir vigorously until all but a few crystals are dissolved.

14. Place the flask at 5° C overnight. During this period the undissolved crystals (mostly leucine and isoleucine) will usually go into solution.

15. Bring the final volume to 1 liter with distilled water, mix well and filter sterilize. Dispense in 10 ml amounts in tubes with rubber lined screw caps. Store at -20° C until ready for use.

16. Just before use dilute stock 1 in 10 and add solution D-G-P (1 ml per 1 liter of 1X medium).

B. EAGLE BASAL MEDIUM (modified for HeLa Cells)

The following formula includes the amino acids and vitamins of Eagle (9-11) medium combined in Hanks BSS rather than Earle salt solution (which requires gassing or addition of buffer for control of pH).

	Material	Amount	Preparation
Unit #1 (100X)	l-Arginine	174.2 mg	Dissolve in glass-distilled water to 100 ml. All of the reagents are soluble.
	l-Histidine	78.0 mg	
	l-Isoleucine	262.0 mg	
	l-Leucine	262.0 mg	
	l-Lysine	292.3 mg	
	l-Phenylalanine	165.2 mg	
	l-Threonine	238.2 mg	
	l-Tryptophane	40.8 mg	
	l-Valine	234.3 mg	
Unit #2 (10X)	l-Cystine	12.0 mg	Dissolve in 100 ml distilled water.
	l-Methionine	7.5 mg	
	l-Tyrosine	18.2 mg	
Unit #3 (100X)	l-Glutamine	2.92 gm	Dissolve in water to 100 ml (easily soluble); sterilize by filtration and store at -20° C.
Unit #4 (100X)	Thiamin-HCl	3.4 mg	Dissolve in 100 ml distilled water. Some opalescence may be present.
	Riboflavin	0.38 mg	
	Ca-pantothenate	4.8 mg	
	Pyridoxal-HCl	2.0 mg	
	Nicotinamide	1.2 mg	
	Choline-Cl	1.4 mg	
	Biotin	2.4 mg	
	i-Inositol (meso)	2.0 mg	
	Folic acid	4.4 mg	

Combine 10 ml of Unit #1, 100 ml of Unit #2, 10 ml of Unit #4 and 370 ml of distilled water. Add to this mixture 500 ml of 2X Hanks solution. Autoclave at 120° C for 15 minutes. Allow solution to cool and aseptically add 10 ml of filter-sterilized glutamine. Adjust to pH 7.2 with sterile 1.4% $NaHCO_3$ solution.

NOTE: Each of the solutions, with the exception of Unit #3 may be autoclaved and stored separately at -20° C.

C. EAGLE MINIMUM ESSENTIAL MEDIUM

In 1959 (12) Eagle revised his original basal medium (9) to give a solution which would permit longer periods of culture of mammalian cells without re-feeding. This medium when supplemented with serum protein is suitable for the cultivation of many cells in either monolayer or suspension. For growing cells in suspension, omission of calcium and a tenfold increase in NaH_2PO_4 is suggested.

This medium may be prepared in the same manner as Eagle Basal Medium (p. 226).

	Material	Amount	
		mM	Mg/liter
	1-arginine	0.6	105
	1-histidine	0.2	31
	1-isoleucine	0.4	52
Unit #1	1-leucine	0.4	52
	1-lysine	0.4	58
	1-phenylalanine	0.2	32
	1-threonine	0.4	48
	1-tryptophane	0.05	10
	1-valine	0.4	46
	1-cystine	0.1	24
Unit #2	1-methionine	0.1	15
	1-tyrosine	0.2	36
Unit #3	1-glutamine	2.0	292
	Thiamine		1
	Riboflavin		0.1
Unit #4	Pantothenate		1
	Pyridoxal		1
	Nicotinamide		1
	Choline		1
	i-inositol		2
	Folic acid		1
Unit #5	Glucose	5.5	1000
	NaCl	116	6800
	KCl	5.4	400
	$CaCl_2$	1.8	200
Unit #6	$MgCl_2 \cdot 6H_2O$	1.0	200
	$NaH_2PO_4 \cdot 2H_2O$	1.1	150
	$NaHCO_3$	23.8	2000

Hanks balanced salt solution may be substituted for solution 5 and 6.

D. TROWELL MEDIUM T8 (Modified)

Stock solution (20X)

	Material	Amount	Preparation
Unit #1	l-Arginine HCl	42 mg	Dissolve tyrosine in 100 ml distilled water heated to 90° C, add remaining amino acids being sure to add tryptophane last when solution has cooled. Sterilize by filtration. Store in 16 x 125 mm screw-cap hard glass tubes. May be kept for six months at room temperature.
	l-Histidine HCl· H_2O	21 mg	
	l-Isoleucine	52 mg	
	l-Leucine	52 mg	
	l-Lysine HCl	72 mg	
	l-Methionine*	15 mg	
	l-Phenylalanine*	33 mg	
	l-Threonine*	48 mg	
	l-Tryptophane	8 mg	
	l-Tyrosine	36 mg	
	l-Valine	46 mg	
Unit #2	NaCl	12. 2 gm	Dissolve in 100 ml distilled water. Sterilize by filtration. Store in 16 x 125 mm screw-cap hard glass tubes. May be kept six months at room temperature.
	KCl	0. 9 gm	
	$CaCl_2$	0. 44 gm	
	$MgSO_4 \cdot 7H_2O$	0. 5 gm	
Unit #3	$NaHCO_3$	5. 64 gm	Dissolve in 100 ml distilled water. Autoclave in 16 x 125 mm screw-cap hard glass tubes at 120° C for 15 minutes. May be stored indefinitely.
	Phenol Red	0. 02 gm	
Unit #4	Insulin**	10 mg	Dissolve ingredients in 10 ml distilled water acidified with 0. 03 ml of 1N HCl. Sterilize by filtration. Store in sterile 16 x 125 mm screw-cap tubes. May be kept one month.
	p-aminobenzoic acid	7 mg	
Unit #5	Glucose	0. 8 gm	Dissolve in 10 ml distilled water. Sterilize by filtration. Make up fresh each time.
	$NaH_2PO_4 \cdot 2H_2O$	90. 0 mg	
	Thiamine HCl	3. 4 mg	
	l-Cysteine HCl	9. 4 mg	

Preparation of Working Trowell Medium (1X)

For use, add 0. 5 ml of each stock solution to 7. 5 ml of distilled water. Equilibrate with 5% CO_2 at 35° C to a pH of 7. 6.

* Trowell (13) provided amounts for the DL forms of these amino acids and therefore the figures provided herein represent half the amount specified by Trowell.

** Eli Lilly and Co., Indianapolis 6, Indiana. Should have a minimum of zinc.

E. WAYMOUTH (14) MB 752/1 MEDIUM

Stock Solutions

	Material	Amount	Preparation
Unit #1	NaCl	1.2 gm	Dissolve all ingredients except buffer salts in 80 ml distilled water. Dissolve buffer salts in small volume distilled and add to above solution. Make up to 100 ml with distilled water. Sterilize by filtration. Store at 5° C. May be kept 3 weeks.
	KCl	30.0 mg	
	$CaCl_2 \cdot 2H_2O$	24.0 mg	
	$MgCl_2 \cdot 6H_2O$	48.0 mg	
	$MgSO_4 \cdot 7H_2O$	40.0 mg	
	Dextrose	1.0 gm	
	Ascorbic Acid	3.5 mg	
	Cysteine HCl	18.0 mg	
	Glutathione	3.0 mg	
	Choline HCl	50.0 mg	
	Hypoxanthine	5.0 mg	
	Glutamine	70.0 mg	
	Na_2HPO_4	60.0 mg	
	KH_2PO_4	16.0 mg	
	$NaHCO_3$	448.0 mg	
Unit #2	Thiamine HCl	40.0 mg	Dissolve ingredients in 100 ml distilled water. Sterilize by filtration. May be kept indefinately.
	Ca Pantothenate	4.0 mg	
	Riboflavin	4.0 mg	
	Pyridoxine HCl	4.0 mg	
	Folic Acid	1.6 mg	
	Biotin	0.08 mg	
	m-inositol$\cdot 2H_2O$	4.0 mg	
	Nicotinamide	4.0 mg	
	Vitamin B_{12}	0.8 mg	
Unit #3	l-lysine HCl	240 mg	Dissolve ingredients in 100 ml distilled water. Adjust pH to 7.4 with NaOH. Sterilize by by filtration. May be stored indefinitely.
	l-histidine HCl	150 mg	
	l-glutamic acid	150 mg	
	l-threonine	75 mg	
	l-arginine HCl	75 mg	
	l-valine	65 mg	
	l-aspartic acid	60 mg	
	Glycine	50 mg	
	l-proline	50 mg	
	l-leucine	50 mg	
	l-methionine	50 mg	
	l-phenylalanine	50 mg	
	l-tyrosine	40 mg	
	l-tryptophan	40 mg	
	l-isoleucine	25 mg	
	l-cystine	15 mg	

(continued on next page)

Waymouth medium 752/1 working solution (1X)

Prepare working solution by adding 37. 5 ml of sterile distilled water, 50 ml of unit #1, 2. 5 ml of unit #2 and 10 ml of unit #3 to give a total of 100 ml.

F. NCTC 109 (15, 16)

Stock Solutions

	Material	Amount	Preparation
Unit #1	l-alanine	26. 8 mg	Dissolve ingredients in 100 ml of 3 x Earle BSS. Sterilize by filtration. Store at 4° C.
	l-α-aminobutyric acid	4. 7 mg	
	l-arginine-HCl	26. 5 mg	
	l-asparagine	6. 9 mg	
	l-aspartic acid	8. 4 mg	
	d-glucosamine-HCl	3. 2 mg	
	l-glutamic acid	7. 0 mg	
	l-glutamine	30. 4 mg	
	Glycine	11. 5 mg	
	l-histidine-HCl	20. 8 mg	
	Hydroxy-l-proline	3. 5 mg	
	l-isoleucine	15. 3 mg	
	l-leucine	17. 4 mg	
	l-lysine-HCl·H_2O	35. 8 mg	
	l-methionine	3. 8 mg	
	l-ornithine-HCl	8. 0 mg	
	l-phenylalanine	14. 0 mg	
	l-proline	5. 2 mg	
	l-serine	9. 1 mg	
	l-taurine	3. 6 mg	
	l-threonine	16. 1 mg	
	l-tryptophan	14. 6 mg	
	l-valine	20. 8 mg	
Unit #2	l-tyrosine	82. 2 mg	Dissolve in a minimum of 0. 075 N HCl and bring to final volume of 50 ml in distilled water. Sterilized by filtration.
	l-cystine	52. 45 mg	

	Material	Amount	Preparation
Unit #3	$\underline{A}_1$		
	Niacin	12.5 mg	Dissolve in boiling distilled water. Bring to a final volume of 25 ml.
	p-aminobenzoic acid	25.0 mg	
	$\underline{A}_2$		
	Niacinamide	12.5 mg	Dissolve in distilled water at room temperature and bring to final volume of 25 ml.
	Pyridoxine-HCl	12.5 mg	
	Pyridoxal-HCl	12.5 mg	
	Thiamine-HCl	5.0 mg	
	d-Calcium pantothenate	5.0 mg	
	i-inositol	25.0 mg	
	Choline chloride	250.0 mg	
	Vitamin B_{12}	2000.0 mg	
	$\underline{A}_3$		
	Riboflavin	10.0 mg	Dissolve in 10 ml of 0.075 N HCl. Add 5 ml of 0.2 N NaOH. Heat gently. Bring to final volume of 20 ml with distilled water.

To prepare Unit #3 A, combine 10 ml of unit A_3 with 25 ml of unit A_1 and 25 ml of unit A_2. Bring to final volume of 100 ml.

	Material	Amount	Preparation
Unit #3	$\underline{B}$		
	d-Biotin	10 mg	Dissolve in 50 ml of distilled water acidified with 1 ml of 1N HCl. Bring to final volume of 100 ml.

	Material	Amount	Preparation
Unit #3	C		
	Folic acid	10 mg	Dissolve in 100 ml of distilled water at about 80° C.
Unit #3	D_1		
	Vitamin D_2 (calciferol) Vitamin A alcohol	10 mg 10 mg	Dissolve the vitamin D in 2 ml of absolute ethanol in a 100 ml volumetric flask. Add the vitamin A to this tincture.
	D_2		
	Menadione (vitamin K)	10 mg	Dissolve in 1 ml of absolute ethanol.
	D_3		
	Tween 80	500 mg	Dissolve in 10 ml distilled water.

To prepare Unit #3 D, add 0.1 ml of unit D_2 and 10 ml of unit D_3 to unit D_1. Bring to final volume of 100 ml with distilled water.

	Material	Amount	Preparation
Unit #3	E		
	Disodium α-tocopherol phosphate	10 mg	Dissolve in 100 ml of distilled water.

To prepare complete Unit #3, combine 2 ml of 3 A with 1 ml of 3 B, 1 ml of 3 C, 10 ml of 3 D and 1 ml of 3 E and add 11 ml distilled water. Sterilize by filtration. Store at 4° C.

	Material	Amount	Preparation
Unit #4	Ascorbic acid Cysteine-HCl Glutathione (monosodium salt)	20.8 mg 108.3 mg 4.2 mg	Dissolve in 25 ml distilled water. Sterilize by filtration. Make up fresh each day.

	Material	Amount	Preparation
Unit #5	Diphosphopyridine nucleotide	2.1 mg	Dissolve in 15 ml of distilled water. Sterilize by filtration. Make up fresh each day.
	Triphosphopyridine nucleotide (Sodium salt) (65% pure)	0.3 mg	
	Coenzyme A (70-75% pure)	0.75 mg	
	Thiamine pyrophosphate (cocarboxylase)	0.3 mg	
	Flavin adenine dinucleotide (58.2% pure)	0.3 mg	
	Uridine triphosphate (90% pure)	0.3 mg	
Unit #6	Deoxyadenosine	3.0 mg	Dissolve in 15 ml of distilled water. Sterilize by filtration. Make up fresh each time.
	Deoxycytidine-HCl	3.0 mg	
	Deoxyguanosine	3.0 mg	
	Thymidine	3.0 mg	
	5-methylcytosine	0.03 mg	
Unit #7	Sodium acetate	100.0 mg	Dissolve in 100 ml distilled water. Sterilize by filtration. Store at -20° C or below.
	Glucuronolactone	3.6 mg	
	Sodium glucuronate	3.6 mg	
	Glutamine	200.0 mg	
Unit #8	Phenol red	5 mg	Dissolve phenol red in 0.5 ml of 0.3 N NaOH. Dilute to 50 ml with 1 x Earle BSS. Sterilize by filtration.

WORKING SOLUTION (1X)

To prepare working solution, combine aseptically 12 ml of unit 1, 1 ml of unit 2, 0.65 ml of unit 3, 6 ml of unit 4, 5 ml of unit 5, 5 ml of unit 6, 5 ml of unit 7, 20 ml of unit 8. Add 1.35 ml of sterile distilled water and 44.0 ml of sterile 1 x Earle BSS.

G. PUCK N16 (6)

Material	gm/liter	Preparation
l-Arginine HCl	0.0375	Dissolve in 1000 ml Saline F (p. 218). Sterilize by filtration.
l-Histidine HCl	0.0375	
l-Lysine HCl	0.080	
l-Tryptophane	0.020	
β-Phenyl-l-alanine	0.025	
l-Methionine	0.025	
l-Threonine	0.0375	
l-Leucine	0.025	
dl-Isoleucine	0.025	
dl-Valine	0.050	
l-Glutamic	0.075	
l-Aspartic	0.030	
l-Proline	0.025	
Glycine	0.100	
Glutamine	0.20	
l-Tyrosine	0.040	
l-Cystine	0.0075	
Hypoxanthine	0.025	
Thiamine HCl	0.0050	
Riboflavin	0.00050	
Pyridoxin HCl	0.00050	
Folic acid	0.00010	
Biotin	0.00010	
Choline	0.0030	
Ca pantothenate	0.0030	
Niacinamide	0.0030	
i-Inositol	0.0010	

To make a complete growth medium, combine 40 ml of N16 (above), 4 ml of NCTC 109 (p. 230), 15 ml of fetal calf serum (pretested for plating efficiency) and 41 ml of Saline F. (p. 218). Antibiotics may be added.

2. MAINTENANCE MEDIA

Maintenance media are designed to permit metabolism of cells, at levels satisfactory for a variety of studies, but do not support cellular proliferation. The widest application of such media is in virus studies. Two general classes of maintenance media are in use: (a) defined media are used when natural materials may contain virus inhibitors or when biochemical studies are being made; (b) natural media or undefined media are often much less expensive and are adequate for many purposes. Innumerable formulae for maintenance media have been published which are excellent for their prescribed purposes.

A. La Ye (Lactalbumin hydrolysate-yeast extract)

An example of an undefined maintenance medium having wide application is La Ye. It is relatively inexpensive and easy to prepare. Lactalbumin hydrolysate provides an abundant source of amino acids while yeast extract contains adequate vitamin complements.

Lactalbumin hydrolysate*	5. 0 gm
Yeast extract**	1. 0 gm
Hanks BSS (without bicarbonate)	1000 ml

Dispense in prescription bottles in desired quantities and sterilize by autoclaving. Before use, add 2 ml of 5% sterile sodium bicarbonate solution per 100 ml.

B. SCHERER MAINTENANCE MEDIUM (17, 18)

Because of its general applicability for cell culture work, a simplified version of Scherer maintenance medium*** is presented. Two groups of compounds have been omitted purposefully.

* Edamin S, Sheffield Chemical, Norwich, New York.
** Difco Laboratories, Detroit, Michigan.
*** Modified from the original formula.

Preparation of Stock Scherer Maintenance Solution (MS-10X)

	Material	Amount	Preparation
Unit #1	Parenamine* (15% with tryptophane)	20.0 ml	Dissolve in 200 ml distilled H_2O.
	Glycine	200.0 mg	
	Histidine (free base, DL)	200.0 mg	
	1-Cystine	150.0 mg	
Unit #2	Glycerol	5.0 gm	Dissolve in 200 ml distilled H_2O.
	Sodium pyruvate	6.4 gm	
	Sodium acetate $\cdot 3H_2O$	11.3 gm	
Unit #3	KH_2PO_4	900.0 mg	Dissolve in 200 ml distilled H_2O.
	Thiamine·HCl	10.0 mg	
	Nicotinamide	4.0 mg	
	Calcium pantothenate (dextrorotary)	4.0 mg	
	Pyridoxal·HCl	4.0 mg	
	Pyridoxamine dihydrochloride	4.0 mg	
	Riboflavin	4.0 mg	
	i-Inositol (meso)	14.0 mg	
	Choline chloride	14.0 mg	
	D-ribose	4.0 mg	
Unit #4	Biotin	1.0 mg	Dissolve in 20 ml distilled H_2O.
	Folic acid	1.0 mg	
	Para aminobenzoic acid	1.0 mg	

Combine Units #1, #2, #3; add 2.0 ml of Unit #4, 100 ml of 0.2% phenol red solution, and dilute to 1 liter with distilled water. Preserve this 10X stock solution with 1-2 ml of chloroform in a reagent bottle.

Unit #5 Hanks BSS, 10X without glucose. (See page 214 for method of preparation.) Conspicuously label the bottle containing BSS without glucose so that it is not used by mistake in other work.

	Material	Amount	Preparation
Unit #6	Glucose	10.0 gm	Dissolve in 100 ml distilled H_2O. Autoclave for 15 minutes at 120° C.

* Winthrop-Stearns, Inc., New York 13, N.Y. or Windsor, Ontario.

	Material	Amount	Preparation
Unit #7	$NaHCO_3$	1. 4 gm	Dissolve and make up to
	Phenol red, 0. 2%	1. 0 ml	100 ml with distilled water;
			autoclave as above.

Preparation of Working Scherer Maintenance Solution (MS)

Add 10 ml each of the 10X MS and BSS stocks to 80 ml of water and autoclave for 15 minutes at 120° C. Cool and aseptically add 2 ml of Unit #6 and 7. 5 ml of Unit #7. Equilibrate to a pH of 7. 6 by storage of the bottles with loosened caps in the refrigerator.

3. GROWTH MEDIA

The choice of growth medium must be governed by several considerations. Among these are (a) the nutritional requirements of the cells employed, (b) requirements of the experiment such as the need for rapid proliferation, differentiation or specialized cell function, (c) need for a chemically defined medium, (d) availability of components, (e) cost.

Most growth media have as their base BSS or a chemically defined medium which is supplemented by one or more natural products, such as serum. The choice of serum must be carefully considered in view of its potential cytotoxicity (19) and of its selective action on clonal variants (20). Several defined media are available which are nutritionally adequate for support of certain cell lines. Choice of one of these defined media will reduce the amount or variety of supplements required.

Listed here are a few examples of growth media. While it is unlikely that any particular medium will be optimal for more than a limited number of cell lines it is wise to standardize, insofar as possible, the media used within a given laboratory.

A. $EAGLE_{80}$ $SERUM_{20}$

A growth medium with wide application may be prepared by supplementing a chemically defined medium such as those of Eagle (p. 226) or Morgan, *et al* (p. 220) with 10-20% whole animal serum. If needed, add penicillin and streptomycin to a final concentration of 100 units/ml and 100 μg/ml respectively.

B. BSS_{60} $SERUM_{40}$

For some cell strains and for the cultivation of monocytes a mixture of serum and BSS is adequate. It is especially important to use homologous serum in such instances as heterologous sera are likely to be toxic at levels above 20%. If needed, add penicillin and streptomycin to a final concentration of 100 units/ml and 100 μg/ml respectively.

C. BSS_{40} $SERUM_{40}$ EE_{20}

This medium is classic in cell culture work and is used commonly for primary explant cultures. Homologous serum should be used (see B., above). Embryo extract is generally regarded as the chief growth stimulating component so that variations in growth response can be obtained by altering the ratio of the three components. The presence of proteinases, deoxyribonuclease, ribonuclease and other enzymes in embryo extract make this medium comparatively unstable. It may be stored at 4° C for 5-7 days or at -20° C for indefinite periods. If needed, add penicillin and streptomycin to a final concentration of 100 units/ml and 100 μg/ml respectively.

D. 199 PEPTONE (199P)

Medium 199 (page 220) may be supplemented with 0.5% Bacto-peptone* to give a medium satisfactory for growth of L-M strain mouse fibroblasts, and other similar cell strains, as monolayers. With the addition of 0.12% methylcellulose** this medium is satisfactory for growth of these cells in suspension culture. If needed, add penicillin and streptomycin to a final concentration of 100 units/ml and 100 μg/ml respectively.

E. YEAST EXTRACT-EAGLE MEDIUM-LACTALBUMIN HYDROLYSATE-PEPTONE (YELP)

This medium is particularly suited for rapid proliferation of L-M strain and other mouse fibroblasts. When supplemented with 5-10% serum it gives excellent growth of many human and animal cells.

* Difco Laboratories, Detroit, Michigan.

** Dow Methocel, 15 cps., Dow Chemical Co., Midland, Michigan.

Material	Amount	Preparation
Yeast extract*	0.5 gm	Dissolve the various ingredients in distilled water and dilute to 1 liter. All ingredients are readily soluble. Sterilize by filtration.
Eagle amino acid stock (100X)**	5.0 ml	
Eagle vitamin stock (100X)**	2.0 ml	
Lactalbumin hydrolysate***	2.5 gm	
Bacto-peptone*	5.0 gm	
Sodium bicarbonate	1.0 gm	
Hanks BSS (10X)	100.0 ml	

If needed, add penicillin and streptomycin to a final concentration of 100 units/ml and 100 µg/ml respectively.

F. EAGLE MEDIUM (2X)

A modified Eagle medium containing twice the amount of amino acids, vitamins and glutamine in a normal Hanks BSS will support continuous growth of L-M mouse cells in monolayer culture (21).

G. EAGLE$_{75}$ TRYPTOSE PHOSPHATE$_{15}$ SERUM$_{10}$

This medium is adequate for growth of most established cell lines. It is made of 75 parts of Eagle medium (p. 226), 15 parts Tryptose phosphate broth**** and 10 parts serum.

* Difco Laboratories, Detroit, Michigan.
** These stocks may be prepared as Units 1-4 on page 226 or obtained from commercial sources as BME (100X).
*** "Edamin S", Sheffield Chemical Co., Norwich, N.Y.
**** Difco Laboratories, Inc., Detroit, Mich. (prepared according to manufacturers instructions, for microorganisms).

MISCELLANEOUS SOLUTIONS

1. CALCIUM AND MAGNESIUM FREE PHOSPHATE BUFFERED SALINE (CMF-PBS)

A Mg++ and Ca++ free salt solution is needed to disperse cells with chelating agents such as EDTA as well as for other purposes. The phosphate buffer is used to maintain pH over short periods of time and does not depend upon shifts in CO_2.

A. 10X STOCK SOLUTION

Material	Amount	Preparation
NaCl	80.0 gm	Dissolve in 1000 ml distilled water. Sterilize by filtration.
KCl	3.0 gm	
Na_2HPO_4	0.73 gm	
KH_2PO_4	0.20 gm	
Glucose	20.0 gm	

B. WORKING SOLUTION

Dilute 1 in 10 with sterile distilled water, dispense in suitable amounts.

2. TRYPSIN

Trypsin used for routine harvesting of cells need not be crystalline.

A. STOCK SOLUTION (4X)

(1) Place 2.0 gm of trypsin* in a beaker. Add 2-3 ml of 1X CMF-PBS (see above) and make into a paste.

(2) Dissolve in 900 ml of CMF-PBS.

(3) Stir continuously until essentially all of the trypsin is dissolved. (A small amount will be insoluble.)

(4) Make up to a final volume of 1000 ml with 1X CMF-PBS and filter through filter paper.

* Trypsin 1:250, Difco Laboratories, Detroit, Michigan.

(5) Sterilize by filtration.

(6) Dispense in convenient volumes and store at -20° C.

B. WORKING SOLUTION (1X)

(1) Dilute stock 1:4 with sterile 1X CMF-PBS to give a final concentration of 0.05% trypsin. It may be desirable to dilute further for some uses.

3. ETHYLENEDIAMINETETRAACETIC ACID (EDTA OR VERSENE*)

EDTA is used for dispersing cells and may be employed as either the disodium or tetrasodium salt. Occasionally the two salts are used in combination to give the desired pH although they have little buffering capacity in the physiological range.

Material	Amount	Preparation
NaCl	8.0 gm	Dissolve in 1000 ml of distilled water. Dispense in convenient amounts and sterilize by autoclaving at 120° C for 15 minutes.
KH_2PO_4	0.2 gm	
KCl	0.2 gm	
Na_2HPO_4	1.15 gm	
EDTA (tetrasodium salt)	0.2 gm	

4. STOCK METHYLCELLULOSE**

The optimal final concentration for suspension culture work is 0.1-0.2%. Due to the peculiar properties of the material it is not practical to prepare a stock solution more concentrated than 4.0%.

(a) Suspend 4 gm of 15 cps. Methocel** in approximately 30 ml of BSS at 90° C. Stir until powder is wetted.

(b) Cool to 4° C and dilute to 100 ml with BSS previously chilled to 4° C. Shake well to dissolve.

(c) Dispense 50 ml amounts into 100 ml screw-cap prescription bottles and autoclave at 120° C for 15 minutes. (The Methocel will boil vigorously during autoclaving and when removed from the sterilizer will have the appearance of coagulated protein.)

(d) Cool to room temperature and refrigerate at 4° C overnight. Methocel will become more fluid as it is chilled.

* A trade name commonly found in the literature.

** Methocel 15 cps., reagent grade, Dow Chemical Co., Midland, Michigan.

5. CITRIC ACID-CRYSTAL VIOLET SOLUTION

Prepare a solution of 0.01% crystal violet in 0.1M citric acid. Dispense the solution in 100 ml amounts and autoclave at 120° C for 15 minutes. (Molds will grow in the citric acid solution if it is not sterilized.)

6. STOCK ANTIBIOTIC SOLUTIONS

Most antibiotics are unstable in solution and should be stored at -20° C. They are then added to medium just before use. It is convenient to prepare and store them as concentrated stock solutions. Nystatin does not dissolve but forms a colloidal suspension. Antibiotics should be prepared aseptically from the sterile commercially available powder by addition of sterile distilled water. Penicillin and streptomycin are conveniently prepared as a single stock solution since they are most frequently used in combination. Antibiotics of the tetracycline series decompose with repeated freezing and thawing.

7. BARIUM SULFATE TURBIDITY STANDARDS (McFarland, 22)

To prepare turbidity standards, mix 1% aqueous barium chloride and 1% sulfuric acid in the following proportions:

	1% barium chloride	1% sulfuric acid
Tube 1	0.1 ml	9.9 ml
Tube 2	0.2 ml	9.8 ml
Tube 3	0.3 ml	9.7 ml
Tube 4	0.4 ml	9.6 ml
Tube 5	0.5 ml	9.5 ml
Tube 6	0.6 ml	9.4 ml
Tube 7	0.7 ml	9.3 ml
Tube 8	0.8 ml	9.2 ml
Tube 9	0.9 ml	9.1 ml
Tube 10	1.0 ml	9.0 ml

Tubes may be sealed and kept for long periods.

8. ALSEVER SOLUTION (MODIFIED) (23)

This solution is used for the preservation of mammalian erythrocytes.

A. Stock Solutions

Material	Amount	Preparation
NaCl $Na_3C_6H_5O_7 \cdot 2H_2O$ (Trisodium citrate) Citric Acid	4.2 gm 8.0 gm 0.55 gm	Dissolve in 900 ml distilled water. Dispense in 90 ml amounts and autoclave.
Glucose (10X)	20.5 gm	Dissolve in 100 ml distilled water. Sterilize by autoclaving.

B. Working Solution

Add 10 ml of 10X glucose stock solution aseptically to 90 ml of sodium citrate solution.

REFERENCES

1. DULBECCO, R., and VOGT, M., Plaque formation and isolation of pure lines with poliomyelitis viruses, J. Exper. Med., 99: 167-182 (1954).
2. EARLE, W. R., Production of malignancy in vitro. IV. The mouse fibroblast cultures and changes seen in living cells, J. Nat. Cancer Inst., 4: 165-212 (1943).
3. HANKS, J. H., The longevity of chick tissue cultures without renewal of medium, J. Cell & Comp. Physiol., 31: 235-260 (1948).
4. HANKS, J. H., and WALLACE, R. E., Relation of oxygen and temperature in the preservation of tissues by refrigeration, Proc. Soc. Exper. Biol. & Med., 71: 196-200 (1949).
5. GEY, G. O., and GEY, M. K., The maintenance of human normal cells and tumor cells in continuous culture. I. Preliminary report: Cultivation of mesoblastic tumors and normal tissue and notes on methods of cultivation. Am. J. Cancer, 27: 45-76 (1936).
6. PUCK, T. T., CIECIURA, S. J., et al., Genetics of somatic mammalian cells. III. Long-term cultivation of euploid cells from human and animal subjects, J. Exper. Med., 108: 945-956 (1958).
7. MORGAN, J. F., MORTON, H. J., et al., Nutrition of animal cells in tissue culture. I. Initial studies on a synthetic medium, Proc. Soc. Exper. Biol. & Med., 73: 1-8 (1950).
8. SALK, J. E., YOUNGNER, J. S., et al., Use of color change of phenol red as the indicator in titrating poliomyelitis virus or its antibody in a tissue-culture system, Amer. J. Hyg., 60: 214-230 (1954).
9. EAGLE, H., Nutrition needs of mammalian cells in tissue culture, Science, 122: 501-504 (1955).
10. EAGLE, H., OYAMA, V. I., et al., Myo-inositol as an essential growth factor for normal and malignant human cells in tissue culture, Science, 123: 845-847 (1956).
11. EAGLE, H., OYAMA, V. I., et al., Myo-inositol as an essential growth factor for normal and malignant human cells in tissue culture, J. Biol. Chem., 226: 191-206 (1957).
12. EAGLE, H., Amino acid metabolism in mammalian cell cultures, Science, 130: 432-437 (1959).
13. TROWELL, O. A., The culture of mature organs in a synthetic medium, Exper. Cell Res., 16: 118-147 (1959).
14. WAYMOUTH, C., Rapid proliferation of sublines of NCTC clone 929 (Strain L) mouse cells in a simple chemically defined medium (MB 752/1). J. Nat. Cancer Inst., 22: 1003-1017 (1959).
15. EVANS, V. J., BRYANT, J. C., et al., Studies of nutrient media for tissue cells in vitro. I. A protein-free chemically defined medium for cultivation of strain L cells, Cancer Res., 16: 77-87 (1956).

15a. EVANS, V. J., BRYANT, J. C., et al, Studies of nutrient media for tissue cells in vitro. II. An improved protein-free chemically defined medium for long-term cultivation of strain L-929 cells, Cancer Res., 16: 88-94 (1956).

16. McQUILKIN, W. T., EVANS, V. J., et al., The adaptation of additional lines of NCTC clone 929 (Strain L) cells to chemically defined protein-free medium NCTC 109. J. Nat. Cancer, Inst., 19: 885-907 (1957).
17. SCHERER, W. F., The utilization of a pure strain of mammalian cells (Earle) for the cultivation of viruses in vitro. I. Multiplication of pseudorabies and herpes simplex viruses, Am. J. Path., 29: 113-137 (1953).
18. SCHERER, W. F., SYVERTON, J. T., et al., Studies on the propagation in vitro of poliomyelitis viruses. IV. Viral multiplication in a stable strain of human malignant cells (Strain HeLa) derived from an epidermal carcinoma of the cervix, J. Exper. Med., 97: 695-710 (1953).
19. WEBB, S. J., and FEDOROFF, S., Natural cytotoxic antibodies in human blood sera which react with mammalian cells and bacteria. II. Effect of heated human serum on microorganisms. Canad. J. Microbiol. 9: 155-162 (1963).
20. MURPHY, W. H., BULLIS, C., et al., Effects of heterologous sera on the modal distribution of variants in four strains of human epithelial cells, Cancer Res., 22: 906-913 (1962).
21. MERCHANT, D. J., and HELLMAN, K. B., Growth of L-M strain mouse cells in a chemically defined medium, Proc. Soc. Exper. Biol. & Med., 110: 194-198 (1962).
22. McFARLAND, J., The nephelometer: An instrument for estimating the number of bacteria in suspensions used for calculating the opsonic index and for vaccines, J. Amer. Med. Assoc., 49: 1176-1178 (1907).
23. BUKANTZ, S. C., REIN, C. R., et al., Studies in complement fixation, II. Preservation of sheep's blood in citrate dextrose mixtures (modified Alsever's solution) for use in the complement fixation reaction. J. Lab. and Clin. Med., 31: 394-399 (1946).

APPENDIX

SOURCES OF SUPPLY

Because of space limitations only a few of the many possible sources of materials are listed. For practical purposes local sources generally are preferred.

Biochemicals and reagents	Aldrich Chemical Co., Inc.	2369 North 29th St. Milwaukee 10, Wisc.
	Antara Chemicals (Div. Gen. Aniline and Film Corp.)	New York, N. Y.
	Armour Pharmaceutical Co. Biochemical Dept.	Kankakee, Illinois
	California Corp. for Biochemical Research	3625 Medford St. Los Angeles 63, Calif.
	Dow Chemical Co.	Midland, Michigan
	Eastman Organic Chemicals	Rochester 3, N. Y.
	Eli Lilly & Co.	Indianapolis, Indiana
	General Biochemicals	Laboratory Park Chagrin Falls, Ohio
	Mann Research Laboratories	136 Liberty St. New York 6, N. Y.
	Nutritional Biochemicals Corp.	Cleveland 28, Ohio
	Pabst Laboratories	1037 W. McKinley Ave. Milwaukee 5, Wisc.

Biochemicals and reagents (con't)	Parke, Davis, Co.	Detroit, Michigan
	Schwartz BioResearch Inc.	230 Washington St. Mt. Vernon, N. Y.
	Sheffield Chemical	Norwich, N. Y.
	Sigma Chemical Co.	3500 Dekalb St. St. Louis 18, Missouri
	Upjohn Co.	Kalamazoo, Michigan
	Worthington Biochemical Corp.	Freehold, N. J.
Cell culture media and reagents	Cappel Laboratories (Div. of Baltimore Biological Laboratories)	West Chester, Pa.
	Colorado Serum Co.	4950 York St. Denver 16, Colorado
	Connaught Medical Research Laboratories	Toronto 4, Ontario, Canada
	Cudahy Laboratories, Inc.	Omaha 7, Nebraska
	Difco Laboratories	920 Henry St. Detroit, Michigan
	Flow Laboratories, Inc.	1710 Chapman Ave. Rockville, Md.
	Grand Island Biological Co.	959 East River Rd. Grand Island, N. Y.
	Hyland Laboratories	4501 Colorado Blvd. Los Angeles 39, Calif.
	Microbiological Associates	4813 Bethesda Ave. Bethesda, Md.
	Pentex, Inc.	P. O. Box 248 Kankakee, Illinois

Cell Lines (Starter cultures)	Dr. R. W. Brown	Carver Foundation Tuskegee Institute Tuskegee, Alabama
	Cell Repository American Type Culture Collection	12301 Parklawn Dr. Rockville, Md. 20852
	Difco Laboratories	920 Henry St. Detroit, Michigan
	Flow Laboratories, Inc.	1710 Chapman Ave. Rockville, Md.
	Microbiological Associates	4813 Bethesda Ave. Bethesda, Md.
	Shamrock Farms, Inc.	Middletown Rd. Middletown, N. Y. 10940
Fermentors and chemostats	American Sterilizer Co.	Erie, Pennsylvania
	Delmar Scientific Laboratories	4701 W. Grand Ave. Chicago 39, Illinois
	Fermentation Design, Inc.	P. O. Box 212 Edison, N. J.
	New Brunswick Scientific Co.	1130 Somerset St. New Brunswick, N. J.
	Stainless Steel Products Co.	17 West 54th St. New York 19, N. Y.
Filters	Ertel Engineering Corp.	Kingston, N. Y.
	Gelman Instrument Co.	600 S. Wagner Rd. Ann Arbor, Michigan
	F. R. Hormann & Co.	17 Stone St. Newark 4, N. J.
	Millipore Filter Corp.	Bedford, Massachusetts
	Selas Corp. of America	Dresher, Pa.

Filters (con't)	Schleicher & Schuell	Keene, N. H.
Rotary action shakers and multiple spinner assemblies	Eberbach Corp.	505 S. Maple Rd. Ann Arbor, Michigan
	Labline, Inc.	Chicago, Illinois
	New Brunswick Scientific Co.	1130 Somerset St. New Brunswick, N. J.
Special tissue culture glassware	Bellco Glass Co.	Vineland, N. J.
	Bonus Laboratories	174 Lowell St. Reading, Mass.
	Corning Glass Works	Corning, N. Y.
	Demuth Glass Works	P. O. Box 629 Parkersburg, W. Va.
	Kimble Glass Co.	Vineland, N. J.
	Kontes Glassware	Vineland, N. J.
	Microchemical Specialties Co.	1825 East Shore Highway Berkeley 10, Calif.
	Quality Glass Apparatus, Inc.	4194 Waters Rd. Ann Arbor, Michigan
	T. C. Wheaton Co.	Millville, N. J.
Special tissue culture plasticware	Falcon Plastics (Div. of Becton-Dickinson Co.)	5500 W. 83rd St. Los Angeles 45, Calif.
Stains and substrates for cytochemistry	California Corporation for Biochemical Research	3625 Medford St. Los Angeles 63, Calif.
	Dajac Laboratories	5000 Longdon St. Philadelphia, Pa.
	General Dyestuff Corp. (Gen. Aniline & Film Corp.)	New York, N. Y.

Stains and substrates for cytochemistry (con't)	Gurr's, Ltd.	London, England
	Matheson, Coleman and Bell or	East Rutherford, N. J. Norwood, Ohio
	National Aniline (Div. of Allied Chem. and Dye Corp.)	40 Rector St. New York 6, N. Y.
	Sigma Chemical Co.	3500 Dekalb St. St. Louis 18, Missouri
	Starkman Biological Laboratories	Toronto, Ontario, Canada
Time lapse cinematographic equipment	C. A. Brinkman & Co.	115 Cutter Mill Road Great Neck, L. I., N. Y.
	Electro Mechanical Development Co.	2337 Bissonnet Houston, Texas
	R. J. Matthias & Associates	2107 DuBarry Dr. Houston, Texas 77018
	Sage Instruments, Inc.	9 Bank St. White Plains, N. Y.
Washing Compounds		
Haemosol	Meinecke & Co.	225 Varick St. New York 14, N. Y.
7X	Linbro Chemicals	New Haven, Conn.
Sud'N	Sepko Chemicals Co.	Minneapolis 21, Minn.
Microsolv	Microbiological Associates	4813 Bethesda Ave. Bethesda 14, Md.
SPECIAL ITEMS		
Coulter Electronic Counter	Coulter Electronics, Inc.	590 W. 20th St. Hialeah, Florida
Culture tube rack and roller drums	Bellco Glass Co. (for Leighton tubes)	Vineland, N. J.

Culture tube rack and roller drums (con't)	Labtool Specialties	5760 Textile Rd. Ypsilanti, Michigan
	Microbiological Associates	4813 Bethesda Ave. Bethesda, Md.
Freezing and Storage		
Ampoules	T. C. Wheaton Co.	Millville, N. J.
Ampoule sealer	Kahlenberg Globe Equipment Co.	P. O. Box 3636 Sarasota, Fla.
	Popper and Sons, Inc.	300 Fourth Ave. New York 10, N. Y.
Canes and boxes for storage	Frozen Semen Products Co. , Inc.	R. D. #1 Breinigsville, Pa.
Dry ice storage chest	Acorn Industries	1425 Rochester Rd. Royal Oak, Michigan
	Canal Industrial Corp.	4940 St. Elmo Ave. Bethesda 14, Md.
Liquid Nitrogen Storage	Air Reduction	150 E. 42nd St. New York 17, N. Y.
	Cryogenic Engineering Co.	200 W. 48th Ave. Denver 16, Colorado
	Linde Division Union Carbide Corp.	270 Park Avenue New York, N. Y. 10017
Mechanical deep-freeze to -90°C.	Revco, Inc.	Deerfield, Michigan
Slow freeze apparatus	Canal Industrial Corp.	4940 St. Elmo Ave. Bethesda 14, Md.
	Linde Division Union Carbide Corp.	270 Park Ave. New York, N. Y. 10017
Heparin (phenol free)	Connaught Medical Research Laboratories	Toronto, Canada

HSR (microscopic mounting medium)	Hartman-Leddon Co.	Philadelphia, Pa.
Observation track for roller tubes	Labtool Specialties	5760 Textile Rd. Ypsilanti, Michigan
	Microbiological Associates	4813 Bethesda Ave. Bethesda, Md.
	Wedco, Inc.	2410 Linden Lane Silver Spring, Md.
"Patapar" glassware wrapping paper	A. J. Buck & Son	1515 E. North Ave. Baltimore, Md.
Propipette	Instrumentation Associates	New York City, N. Y.
Stainless Steel wire mesh	Newark Wire Cloth Co.	351 Verona Ave. Newark, N. J.
	United Foundation Wire Mesh United Surgical Supply Co.	154 Midland Ave. Port Chester, N. Y.
Vinyl cup panels	Fabri-Kal Corporation	242 E. Kalamazoo Ave. Kalamazoo, Michigan
White rubber stopper and silicone stoppers	West & Company	Phoenixville, Pa.

TISSUE CULTURE LITERATURE

Because of the wide range of applications of cell and organ culture the literature concerning these techniques is scattered through many journals. The list given here is not intended to be complete. However, the journals listed contain perhaps 75% of the articles published currently in English.

Acta Pathologica et Microbiologica Scandanavica
American Journal of Clinical Pathology
American Journal of Pathology
Archives fur die Gesamte Virusforschung
Archives of Biochemistry
Anatomical Record
Biochemical and Biophysical Research Communications
Biochemica et Biophysicà Acta
Biochemical Journal
Blood
British Journal of Experimental Pathology
Canadian Journal of Biochemistry and Physiology
Cancer Research
Development Biology
Experimental Cell Research
Gann
In Vitro
Japanese Journal of Experimental Medicine
Journal of Bacteriology
Journal of Biophysical and Biochemical Cytology
Journal of Biological Chemistry
Journal of Cell Biology
Journal of Cellular and Comparative Physiology
Journal of Embryology and Experimental Morphology
Journal of Experimental Medicine
Journal of Experimental Zoology
Journal of Immunology
Journal of Laboratory and Clinical Medicine
Journal of the National Cancer Institute
Journal of Pathology and Bacteriology
Nature
National Academy of Science, Proceedings
New York Academy of Science, Transactions and Annals
Proceedings of the Society for Experimental Biology and Medicine
Radiation Research
Science
Texas Reports on Biology and Medicine
Virology

LOG TO BASE 2

Graphic representation of the growth of cells usually is achieved best by using a logarithmic scale. Calculations are more readily interpreted if logarithms to the base 2 are used instead of common logarithms (*e. g.* log to base 10) for an increase of one unit on the logarithm scale then corresponds to the equivalent of one division by each cell.

The following table shows the logarithm of every integer from 100 to 999. Thus, by direct reading, $\log_2 673 = 9.394$. Similarly, if the highest and lowest count of a series does not differ by more than a factor of 10, there may be no necessity to use logarithms outside of the range of the table. For example, if the cell counts range from 1.7 to 8.9, expression of the count in terms of a volume unit 100 times that of the standard will give values ranging from 170 to 890; all logarithms can then be read directly from the table and graphed.

If the cell counts extend over a wider range, logarithms of numbers not in the table will be needed. These are obtained by the rule that the <u>logarithm of a product or quotient is the sum or difference of the logarithms of the two parts</u>. Thus:

$$\log_2 2692 = \log_2 4 + \log_2 673$$
$$= 2.000 + 9.394 = 11.394$$

$$\log_2 0.075 = \log_2 300 - \log_2 400 - \log_2 10 \text{ (Since } 0.075 = 3/4 \div 10)$$
$$= 8.229 - 8.644 - 3.322 = -3.737$$

The measurements rarely will be accurate to more than three significant digits, but, if required, logarithms of quantities intermediate between tabular values can be derived by interpolation. Thus, $\log_2 2692$ can be calculated alternatively as 2/10 of the amount between:

$$\log_2 2690 = \log_2 269 + \log_2 10$$
$$= 8.071 + 3.222 \qquad = 11.393$$

and $$\log_2 2700 = \log_2 270 + \log_2 10$$
$$= 8.077 + 3.222 \qquad = 11.399$$

and is 11.394 as above.

When suitable factors are not obvious, powers of 10 can be used, remembering that $\log_2 10 = 3.3219$, $\log_2 100 = 6.6439$, $\log_2 1000 = 9.9568$. Logarithms of powers of 10 up to 10^{10} are appended to the table. Hence:

$$\log_2 0.875 = \log_2 875 - \log_2 1000$$
$$= 9.773 - 9.966 = -0.193$$

Of course, for any particular number, multiplication of its common logarithms by 3.3219 ($-\log_2 10$) gives the logarithm to base 2.

LOGARITHMS TO BASE 2

	0	1	2	3	4	5	6	7	8	9
100	6.644	6.658	6.672	6.687	6.700	6.714	6.728	6.741	6.755	6.768
110	.781	.794	.807	.820	.833	.845	.858	.870	.883	.895
120	.907	.919	.931	.943	.954	.966	.977	.989	7.000	7.011
130	7.022	7.033	7.044	7.055	7.066	7.077	7.087	7.098	.109	.119
140	.129	.140	.150	.160	.170	.180	.190	.200	.209	.219
150	7.229	7.238	7.248	7.257	7.267	7.276	7.285	7.295	7.304	7.313
160	.322	.331	.340	.349	.358	.366	.375	.384	.392	.401
170	.409	.418	.426	.435	.443	.451	.459	.468	.476	.484
180	.492	.500	.508	.516	.524	.531	.539	.547	.555	.562
190	.570	.577	.585	.592	.600	.607	.615	.622	.629	.637
200	7.644	7.651	7.658	7.665	7.672	7.679	7.687	7.693	7.700	7.707
210	.714	.721	.728	.735	.741	.748	.755	.762	.768	.775
220	.781	.788	.794	.801	.807	.814	.820	.827	.833	.839
230	.845	.852	.858	.864	.870	.877	.883	.889	.895	.901
240	.907	.913	.919	.925	.931	.937	.943	.948	.954	.960
250	7.966	7.972	7.977	7.983	7.989	7.994	8.000	8.006	8.011	8.017
260	8.022	8.028	8.033	8.039	8.044	8.050	.055	.061	.066	.071
270	.077	.082	.087	.093	.098	.103	.109	.114	.119	.124
280	.129	.134	.140	.145	.150	.155	.160	.165	.170	.175
290	.180	.185	.190	.195	.200	.205	.209	.214	.219	.224
300	8.229	8.234	8.238	8.243	8.248	8.253	8.257	8.262	8.267	8.271
310	.276	.281	.285	.290	.295	.299	.304	.308	.313	.317
320	.322	.326	.331	.335	.340	.344	.349	.353	.358	.362
330	.366	.371	.375	.379	.384	.388	.392	.397	.401	.405
340	.409	.414	.418	.422	.426	.430	.435	.439	.443	.447
350	8.451	8.455	8.459	8.464	8.468	8.472	8.476	8.480	8.484	8.488
360	.492	.496	.500	.504	.508	.512	.516	.520	.524	.527
370	.531	.535	.539	.543	.547	.551	.555	.558	.562	.566
380	.570	.574	.577	.581	.585	.589	.592	.596	.600	.604
390	.607	.611	.615	.618	.622	.626	.629	.633	.637	.640

	0	1	2	3	4	5	6	7	8	9
400	8.644	8.647	8.651	8.655	8.658	8.662	8.665	8.669	8.672	8.676
410	.679	.683	.687	.690	.693	.697	.700	.704	.707	.711
420	.714	.718	.721	.725	.728	.731	.735	.738	.741	.745
430	.748	.752	.755	.758	.762	.765	.768	.771	.775	.778
440	.781	.785	.788	.791	.794	.798	.801	.804	.807	.811
450	8.814	8.817	8.820	8.823	8.827	8.830	8.833	8.836	8.839	8.842
460	.845	.849	.852	.855	.858	.861	.864	.867	.870	.873
470	.877	.880	.883	.886	.889	.892	.895	.898	.901	.904
480	.907	.910	.913	.916	.919	.922	.925	.928	.931	.934
490	.937	.940	.943	.945	.948	.951	.954	.957	.960	.963
500	8.966	8.969	8.972	8.974	8.977	8.980	8.983	8.986	8.989	8.992
510	.994	.997	9.000	9.003	9.006	9.008	9.011	9.014	9.017	9.020
520	9.022	9.025	.028	.031	.033	.036	.039	.042	.044	.047
530	.050	.053	.055	.058	.061	.063	.066	.069	.071	.074
540	.077	.079	.082	.085	.087	.090	.093	.095	.098	.101
550	9.103	9.106	9.109	9.111	9.114	9.116	9.119	9.122	9.124	9.127
560	.129	.132	.134	.137	.140	.142	.145	.147	.150	.152
570	.155	.157	.160	.162	.165	.167	.170	.172	.175	.177
580	.180	.182	.185	.187	.190	.192	.195	.197	.200	.202
590	.205	.207	.210	.212	.214	.217	.219	.222	.224	.226
600	9.229	9.231	9.234	9.236	9.238	9.241	9.243	9.246	9.248	9.250
610	.253	.255	.257	.260	.262	.264	.267	.269	.271	.274
620	.276	.278	.281	.283	.285	.288	.290	.292	.295	.297
630	.299	.301	.304	.306	.308	.311	.313	.315	.317	.320
640	.322	.324	.326	.329	.331	.333	.335	.338	.340	.342
650	9.344	9.347	9.349	9.351	9.353	9.355	9.358	9.360	9.362	9.364
660	.366	.369	.371	.373	.375	.377	.379	.382	.384	.386
670	.388	.390	.392	.394	.397	.399	.401	.403	.405	.407
680	.409	.412	.414	.416	.418	.420	.422	.424	.426	.428
690	.430	.433	.435	.437	.439	.441	.443	.445	.447	.449
700	9.451	9.453	9.455	9.457	9.459	9.461	9.464	9.466	9.468	9.470
710	.472	.474	.476	.478	.480	.482	.484	.486	.488	.490
720	.492	.494	.496	.498	.500	.502	.504	.506	.508	.510
730	.512	.514	.516	.518	.520	.522	.524	.526	.527	.529
740	.531	.533	.535	.537	.539	.541	.543	.545	.547	.549
750	9.551	9.553	9.555	9.557	9.558	9.560	9.562	9.564	9.566	9.568
760	.570	.572	.574	.576	.577	.579	.581	.583	.585	.587
770	.589	.591	.592	.594	.596	.598	.600	.602	.604	.605
780	.607	.609	.611	.613	.615	.617	.618	.620	.622	.624
790	.626	.628	.629	.631	.633	.635	.637	.638	.640	.642
800	9.644	9.646	9.647	9.649	9.651	9.653	9.655	9.656	9.658	9.660
810	.662	.664	.665	.667	.669	.671	.672	.674	.676	.678
820	.679	.681	.683	.685	.687	.688	.690	.692	.693	.695
830	.697	.699	.700	.702	.704	.706	.707	.709	.711	.713
840	.714	.716	.718	.719	.721	.723	.725	.726	.728	.730

	0	1	2	3	4	5	6	7	8	9
850	9.731	9.733	9.735	9.736	9.738	9.740	9.741	9.743	9.745	9.747
860	.748	.750	.752	.753	.755	.757	.758	.760	.762	.763
870	.765	.767	.768	.770	.771	.773	.775	.776	.778	.780
880	.781	.783	.785	.786	.788	.790	.791	.793	.794	.796
890	.798	.799	.801	.803	.804	.806	.807	.809	.811	.812
900	9.814	9.815	9.817	9.819	9.820	9.822	9.823	9.825	9.827	9.828
910	.830	.831	.833	.834	.836	.838	.839	.841	.842	.844
920	.845	.847	.849	.850	.852	.853	.855	.856	.858	.860
930	.861	.863	.864	.866	.867	.869	.870	.872	.873	.875
940	.877	.878	.880	.881	.883	.884	.886	.887	.889	.890
950	9.892	9.893	9.895	9.896	9.898	9.899	9.901	9.902	9.904	9.905
960	.907	.908	.910	.911	.913	.914	.916	.917	.919	.920
970	.922	.923	.925	.926	.928	.929	.931	.932	.934	.935
980	.937	.938	.940	.941	.943	.944	.945	.947	.948	.950
990	.951	.953	.954	.956	.957	.959	.960	.961	.963	.964

10	10^2	10^3	10^4	10^5	10^6	10^7	10^8	10^9	10^{10}
3.322	6.644	9.966	13.288	16.610	19.932	23.253	26.575	29.897	33.219

FINLEY, D. J., HAZLEWOOD, T., et al., Logarithms to Base 2, J. Gen Microbiol., 12: 222-225 (1955).

GLOSSARY

AMEBOID - An adjective used to describe cells characterized by a high degree of motility and with little or no tendency to grow as colonies. Also used to describe monolayer cultures of such cells.

BSS - Balanced salt solution. A buffered isotonic and iso-osmolar salt solution with the mono- and di-valent cations carefully balanced.

CAMERA LUCIDA - A prismatic ocular used to project an image from a microscope onto a piece of paper so that its outline may be traced.

CENTROMERE - A specialized region of the chromosome. This region possesses different staining properties from the rest of the chromosome and represents the site of spindle attachment. The position of the centromere is specific for any one chromosome and is useful in identifying them.

CLEAR - A technical step in the preparation of microscopic slides which follows dehydration, wherein the cells or tissues are treated with a liquid of high refraction. This treatment makes the objects more transparent.

CMF-PBS - Calcium and magnesium-free phosphate buffered salt solution.

CPE - See cytopathic effect.

CYTOPATHIC EFFECT - The morphological changes which cells may demonstrate in response to viral or toxic agents. The phenomenon may range from simple foaming of the cytoplasm or focal clumping of cells to complete destruction.

CYTOTOXIC ENDPOINT - An arbitrary endpoint, based on changes in cell morphology, used to define the effects of viruses, chemicals, *etc.*

DEHYDRATION - A technical step in the preparation of microscopic slides wherein the cells or tissues are passed through a series of alcoholic baths of increasing concentration to remove water.

DIFFERENTIATION - The acquisition of a function or a character which is distinctive or specialized.

EE - Embryo extract. A saline extract of embryos used as a component of culture medium and to initiate clotting of plasma.

FIXATIVE - A fluid which will rapidly kill cells or tissues and prepare them for subsequent staining and observation.

GOLGI APPARATUS - A cytoplasmic organelle which usually appears as a lamellar and vacuolated area near the cell center and is thought to be associated with secretion.

GROWTH - An increase in the mass of living substance and/or the number of cells.

IDIOGRAM - A diagramatic representation of the karyotype which may be based on measurement of the chromosomes in several or in many cells.

INDUCTION - The stimulating and directing effect shown by certain tissues on neighboring tissues or parts in early embryogenesis.

IN SITU - In its natural place or location.

IN VITRO - Within a test tube or culture vessel.

IN VIVO - Within the living animal.

KARYOTYPE - A systematized array of the chromosomes of a single cell prepared either by drawing or photography. In some instances the karyotype of a single cell can typify the chromosomes of an individual or even a species.

KINETOCHORE - See centromere.

LaYe - A medium containing lactalbumin hydrolysate and yeast extract. Used primarily as a maintenance medium.

LETHAL ENDPOINT - An arbitrary endpoint used to describe the effect of viruses, chemical agents, *etc.* on cells. It is dependent upon the subjective determination of cell death based upon vital staining, failure to reproduce, loss of metabolic function or other similar methods.

MECHANOCYTE - A synonym for fibroblastic cells.

METACENTRIC - An adjective to describe a chromosome whose centromere is located at the center of the chromosome.

MITOCHONDRIA - Motile cytoplasmic organelles found in all cells which vary in shape from small spheres to filamentous rods. These structures represent the primary site of oxidative phosphorylation. May be stained vitally with Janus Green B.

MITOTIC COEFFICIENT - The proportion of cells in a population which are undergoing division at a given moment of time.

MITOTIC INDEX - The proportion of cells in a population undergoing division within a given unit of time.

PATCHING - A procedure which involves the removal of lysed plasma from a clot and renewal with fresh plasma.

PBS - Phosphate buffered physiological saline.

PCV - Packed cell volume. A measure of total cell mass.

PFU - Plaque forming unit. A unit of measure for virus content. The smallest quantity of a virus-containing fluid which will give rise to a "plaque" or localized CPE in a monolayer culture under defined conditions.

PHYTOHEMAGGLUTININ - One of a group of compounds obtained from plant sources which is characterized by the ability to bring about agglutination of erythrocytes. Many of these compounds are also potent agents for the stimulation of mitotic activity.

PLAQUE - A localized area or cellular destruction in a monolayer culture.

PPLO - Pleuropneumonia-like organisms. A group of microorganisms belonging to the Genus Mycoplasma. This term is often used loosely by workers in cell culture to include L-forms of a variety of bacterial species.

RAFT - A structural support for an explant, as in organ culture.

SUBMETACENTRIC - An adjective to describe a chromosome whose centromere is located near the center of the chromosome.

SUBTELOCENTRIC - An adjective to describe a chromosome whose centromere is located near the end of the chromosome.

SUPRAVITAL STAINING - Selective staining of living cells.

TCID - Tissue culture infective dose. A unit of measure for virus content. The smallest amount of a virus-containing fluid which will infect a cell culture under defined conditions.

TELOCENTRIC - An adjective to describe a chromosome whose centromere is located at the end of the chromosome.

VERSENE - Trade name for ethylenediaminetetraacetic acid.

YELP - A growth medium consisting of Eagle basal medium supplemented with lactalbumin hydrolysate, yeast extract and peptone.

PROPOSED USAGE OF ANIMAL TISSUE CULTURE TERMS*

The field of Tissue Culture has developed very rapidly during recent years and tissue culture methodology has become a useful research tool in many branches of science. Consequently, scientists with various backgrounds and various aims have entered the field of Tissue Culture. As a result considerable controversy has developed over terminology, and this has hampered communications among scientists in the field. Therefore, in 1961 the Tissue Culture Association formed an Ad Hoc Committee on Nomenclature under the chairmanship of Dr. R. S. Chang, of Harvard University. This committee studied the problem very thoroughly and after three years of deliberation prepared a report on "Tissue Culture Terminology." This report accomplished a very important task; it consolidated thinking in some areas, and pinpointed areas where considerable disagreement still existed.

In 1964 a smaller committee was set up, consisting of Dr. V. J. Evans, National Cancer Institute, Bethesda, Dr. T. C. Hsu, M. D. Anderson Hospital, Houston and Dr. S. Fedoroff (Chairman), University of Saskatchewan. It was charged with these duties: to further study the problem, to analyze the reasons for controversy on some terms, and to propose a generally acceptable terminology. This committee prepared a report on terminology and by means of thorough discussion with various groups of scientists, the report underwent five major revisions. The final, fifth draft of a proposed usage of animal tissue culture terms was accepted by the Tissue Culture Association at its Annual Meeting on June 3, 1966 in San Francisco. It also received the approval of a number of workers in the field from abroad.

The adopted text of the "Proposed Usage of Animal Tissue Culture Terms" follows.

Tissue Culture

Animal Tissue Culture is concerned with the study of cells, tissues and organs explanted from animals and maintained or grown *in vitro* for more than 24 hours. Dependent upon whether cells, tissues or organs are to be maintained or grown, two methodological approaches have been developed in the field of Tissue Culture.

(a) Cell culture. This term is used to denote the growing of cells *in vitro* including the culture of single cells. In cell cultures the cells are no longer organized into tissues.

* Published with the permission of the Tissue Culture Association.

(b) Tissue or Organ culture. This term denotes the maintenance or growth of tissues,* organ primordia, or the whole or parts of an organ *in vitro* in a way that may allow differentiation and preservation of the architecture and/or function.

Explant

This term describes an excised fragment of a tissue or an organ used to initiate an *in vitro* culture.

Monolayer

This term refers to a single layer of cells growing on a surface.

Suspension culture

This term denotes a type of culture in which cells multiply while suspended in medium.

Primary culture

This term implies a culture started from cells, tissues or organs taken directly from organisms. It does not include cultures started from explants of tumors developed by injecting cultured cells into animals. Such cultures would be considered more properly as continuations of the injected cell line or strain. A primary culture may be regarded as such until it is subcultured for the first time. It then becomes a cell line.

Cell line

A "cell line" arises from a primary culture at the time of the first subculture. The term "cell line" implies that cultures from it consist of numerous lineages of the cells originally present in the primary culture. (See note following section on "sub-strain.")

Established cell line

A cell line may be said to have become "established" when it demonstrates the potential to be subcultured indefinitely *in vitro*.

NOTE: With present knowledge about cell behavior *in vitro* it is really impossible to determine the point at which a culture has become "established," if the term "established" is taken to mean that the culture has

*The word "tissue" comes from a French word meaning "woven together" and a tissue in the biological sense is the woven mass of cells together with their intercellular substance. Histologists generally recognize four basic tissues: epithelial, connective, muscle and nervous tissues.

acquired the ability to grow indefinitely. Based only on experience with human fibroblast-like cells, a culture must be subcultured at least 70 times, with intervals of 3 days between subcultures, before it can be considered to be "established." As new information becomes available, it will be necessary to determine for each species the number of transplantations and conditions required to give rise to established cell lines.

Cell strain

A "cell strain" can be derived either from a primary culture or a cell line by the selection or cloning of cells having specific properties or markers. The properties or markers must persist during subsequent cultivation. In describing a cell strain its specific feature should be defined, *e.g.* a cell strain with a certain marker chromosome, a cell strain having a specific antigen, *etc.* (See note following section on "sub-strain.")

Sub-strain

A "sub-strain" can be derived from a strain by isolating a single cell or groups of cells having properties or markers not shared by all cells of the strain.

NOTE: In everyday usage "cell line" and "cell strain" have sometimes been considered interchangeable. However, according to Roget's Thesaurus the terms are not synonymous. The word "line" is used most properly with the meaning of an uninterrupted sequence. In this context it is synonymous with such words as procession, series and succession. In relation to paternity, "line" is synonymous with tree, pedigree, descent and family. In relation to posterity, "line" is synonymous with straight descent. "Strain" is generally used with the meaning of relationships of kindred. In this context "strain" is synonymous with race, stock, clan and tribe. Therefore it is proposed that the term "cell line" in the field of Tissue Culture should imply an uninterrupted sequence of cell growth, whereas the term "cell strain" should imply a certain relationship of the cells, *i.e.* that they all have one or more common properties or markers for which these cells were specifically selected. (See "Cell strain.")

Clone

This term denotes a population of cells derived from a single cell by mitoses. A clone is not necessarily homogeneous and therefore the terms "clone" or "cloned" should not be used to indicate homogeneity in a cell population.

Cloned strain or line

This term denotes a strain or line descended directly from a clone. (See "Clone.")

Diploid cell line

This term denotes a cell line in which, arbitrarily, at least 75 per cent of the cells have the same karyotype as the normal cells of the species from which the cells were originally obtained. It should be noted that a diploid chromosome number is not necessarily equivalent to the diploid karyotype,* because there are situations in which a cell may lose one type of chromosome and acquire another type of chromosome. Thus the karyotype of the cell has changed but the diploid number of chromosomes remains the same. Such cells should be referred to as "pseudo-diploid."

A description of a diploid cell line should include the actual numbers of cells examined, the percentage of diploid cells and their karyotype.

Heteroploid cell line

This term denotes a cell line having less than 75 per cent of cells with diploid chromosome constitution. This term does not imply that the cells are malignant or that they are able to grow indefinitely *in vitro*. In describing a heteroploid cell line, in addition to the karyotype of the stem line the percentage of cells with such karyotype should be stated.

NOTE: For more precise definition of the genetic state of the cells, according to A. Levan and A. Müntzing (Terminology of Chromosome Numbers. Portugaliae Acta Biologica, 7: 1-16, 1963) the following terms should be used:

Haploid (1) The basic number of a polyploid series (symbol: x). Haploid in this meaning = monoploid.

(2) The chromosome number of the haplophase, the gametic, reduced number (symbol: n).

Diploid, triploid, tetraploid, *etc.*

The double, triple, quadruple, *etc.* basic number (symbols: 2x, 3x, 4x, *etc.*).

*The term "Karyotype" has been defined as "... a systematized array of the chromosomes of a single cell prepared either by drawing or by photography, with the extension in meaning that the chromosomes of a single cell can typify the chromosomes of an individual or even a species. The term idiogram would then be reserved for the diagrammatic representation of a karyotype, which may be based on measurements of the chromosomes in several or many cells." A. Robinson, A Proposed Standard System of Nomenclature of Human Mitotic Chromosomes. J.A.M.A., 174: 159, 1960.

Polyploid

General designation for multiples of the basic number, higher than diploid.

Heteroploid

(1) In organisms with predominating diplophase: all chromosome numbers deviating from the normal chromosome number of the diplophase.

(2) In organisms with predominating haplophase: all chromosome numbers deviating from the normal chromosome number of the haplophase.

Euploid

All exact multiples of x.

Aneuploid

All numbers deviating from x and from exact multiples of x.

Mixoploidy

The presence of more than one chromosome number in a cellular population.

Endopolyploidy

The occurence in a cellular population of polyploid cells, which have originated by endomitosis.

Subculture

This term denotes the transplantation of cells from one culture vessel to another.

Passage

This term is synonymous with "subculture," and can denote the passage of cells from one culture vessel to another. When this term is used, however, it must be made clear just what is being passed, because virologists use this term to denote the transfer of supernatant culture fluid rather than cells.

Subculture number

This term indicates the number of times cells have been subcultured, *i.e.*, transplanted from one culture vessel to another.

Subculture interval

This term denotes the interval between subsequent subcultures of cells. This term has no relationship to the term "cell generation time."

Cell generation time

This term denotes the interval between consecutive divisions of a cell. This interval can be best determined at present with the aid of cinematography. This term is not synonymous with "Population doubling time."

Population doubling time

This term is used when referring to an entire population of cells and indicates the interval in which, for example, 1×10^6 cells increase to 2×10^6 cells. This term is not synonymous with "Cell generation time."

Absolute plating efficiency

This term indicates the percentage of individual cells which give rise to colonies when inoculated into culture vessels. The total number of cells in the inoculum, type of culture vessel and the environmental conditions (medium, temperature, closed or open system, CO_2 atmosphere, *etc.*) should always be stated.

Relative plating efficiency

This term indicates the percentage of inoculated cells which gives rise to colonies, relative to a control in which the absolute plating efficiency is arbitrarily set as 100 per cent. The total number of cells in the inoculum, the environmental conditions, and the absolute plating efficiency of the control should always be stated.

Fibroblasts

Fibroblasts are cells of spindle or irregular shape, and as their name implies, are responsible for fiber formation. In cell cultures many other cell types are morphologically indistinguishable from fibroblasts. In organ and tissue cultures in which cell inter-relationships are preserved, fibroblasts may be identified using accepted histological criteria.

Fibroblast-like cells

In cell cultures various types of cells acquire similar morphology. Cells acquiring irregular or spindle shape are often referred to as "fibroblasts." However, the derivation of these cells or their potentialities, such as production of fibers, are usually not known. Therefore such cells are more properly called "fibroblast-like" cells.

Epithelial cells

This term refers to cells apposed to each other forming continuous mosaic-like sheets with very little intercellular substance, as seen *in vivo* or in tissue or organ cultures.

Epithelial-like cells

In cell cultures epithelial cells may assume various shapes but tend to form sheets of closely adherent polygonal cells. However, the degree of cohesion of the cells can vary. When the only criterion for identification of such cells is their tendency to adhere to each other it is preferable to refer to the cells as "epithelial-like" cells.

Culture alteration

This term is used to indicate a persistent change in the properties or behavior of a culture, *e.g.* altered morphology, chromosome constitution, virus susceptibility, nutritional requirements, proliferative capacity, malignant characters, *etc.* The term should always be qualified by a precise description of the change which has occurred in the culture. The term "cell transformation" should be reserved to mean changes induced in the cells by the introduction of new genetic material. The nature and source of the genetic material inducing the change should be specified.

Tumorigenicity

There are no criteria for determining the tumorigenicity of cells observed only *in vitro*. The tumorigenicity of cells can be determined by the behavior of the cells in animals. Whenever, therefore, tumorigenicity is referred to, the following information should be given:

1. Description of host injected (species, strain, sex, age, *etc.*).

2. Pre-treatment, if any, of host.

3. Site of injection and number of cells injected.

4. Description of growth of the injected cells.

5. Histological description of tumor and relation of tumor to adjacent normal tissues.

6. Occurrence of metastases.

Toxicity

The term "toxicity," when used to describe an effect observed in cultures, is meaningful only when the effect itself is also described, *e.g.* toxicity as evidenced by altered morphology of the cells, by failure of the cells to attach to surfaces, by changes in the rate of cell growth, cell death, *etc.*

In reporting a new cell line, the recommendations of the 1957 International Tissue Culture Meeting in Glasgow should be followed. It was recommended that authors should give the following information when first mentioning a cell line in the course of a publication:

1. Whether the tissue of origin was normal or neoplastic and, if neoplastic, whether benign or malignant;

2. Whether the tissue was adult or embryonic;

3. The animal species of origin;

4. The organ of origin;

5. The cell type (if known);

6. The designation of the line;

7. Whether the line has been cloned.

It was further suggested that the designation of the line should consist of a series of not more than four letters indicating the laboratory of origin, followed by a series of numbers indicating the line, *e.g.* NCL 123.

In describing a cell line as much information about the cells as is known should be given. In characterizing a cell line it is advisable to follow the procedures developed by the Cell Culture Collection Committee and adopted by the American Type Culture Collection Cell Repository. The following information may be included in whole or in part:

1. History